Contents

Contributors   ix

Abbreviations   xi

Preface   xvii

1 What is gene therapy?   1
J. Drew and L.-A. Martin
Introduction   1
Germline gene therapy   1
Somatic gene therapy   2
Viral systems   4
Nonviral systems   6
Candidate diseases for application of gene therapy   8
Summary   9
References   9
Further reading   10

2 Why gene therapy?   11
L.-A. Martin and J. Drew
Introduction   11
What makes a disease suitable for gene therapy?   11
Cystic fibrosis   12
Adenosine deaminase deficiency   13
Familial hypercholesterolemia   13
Cancer – could gene therapy provide a cure?   14
Infectious diseases   16
Ethical and regulatory considerations   17
Gene therapy industry – an explosion waiting to happen?   17
Future prospects for gene therapy   19
Further reading   19

3 Viral delivery systems for gene therapy   **21**
S.J. Murphy
Introduction   21
Retrovirus-based vectors   22
Adenovirus-based vectors   26
Adeno-associated virus-based vectors   33
Herpes simplex virus-based vectors   34
Other viral vectors and hybrid vector systems   35
References   38

4 Nonviral delivery systems for gene therapy   **43**
A.D. Miller
Introduction   43
Cationic liposome/micelle-based nonviral delivery systems   45
Cationic polymer-based nonviral delivery systems   52
Alternative chemical nonviral delivery systems   57
Physical nonviral delivery systems   58
Future prospects for nonviral delivery systems   59
References   60

5 Gene therapy for monogenic diseases   **71**
G. Vassaux
Introduction   71
Gene therapy for adenosine deaminase deficiency   71
Gene therapy for cystic fibrosis   78
Conclusion   82
Further reading   82

6 Gene therapy for multifactorial genetic disorders   **83**
A.S. Rigg
Introduction   83
Colorectal cancer   83
Atherosclerotic vascular diseases   92
Diabetes mellitus   94
Summary   97
Further reading   98

7 Gene therapy for infectious diseases   **99**
S. Fidler and J. Weber
Introduction   99
Strategies for antiviral gene therapy   100
The use of gene therapy to manipulate the immune response   113
Conclusions   122
References   122

Understanding Gene Therapy

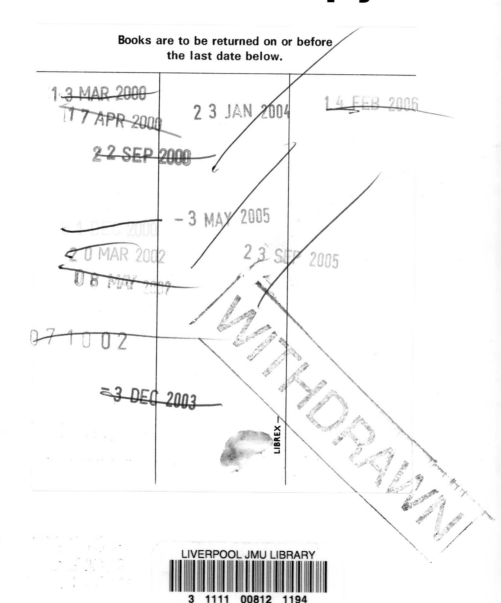

THE MEDICAL PERSPECTIVES SERIES

Advisors:

B. David Hames *School of Biochemistry and Molecular Biology, University of Leeds, UK.*

David R. Harper *Department of Virology, Medical College of St Bartholomew's Hospital, London, UK.*

Andrew P. Read *Department of Medical Genetics, University of Manchester, Manchester, UK.*

Oncogenes and Tumor Suppressor Genes
Cytokines
The Human Genome
Autoimmunity
Genetic Engineering
Asthma
HIV and AIDS
Human Vaccines and Vaccination
Antibody Therapy
Antimicrobial Drug Action
Molecular Biology of Cancer
Antiviral Therapy
Molecular Virology *Second Edition*
DNA Fingerprinting *Second Edition*
Understanding Gene Therapy

Forthcoming titles:

Molecular Diagnosis
Genetic Engineering *Second Edition*

Understanding
Gene Therapy

Edited by

N.R. Lemoine
ICRF Molecular Oncology Unit,
Imperial College School of Medicine, Hammersmith Campus,
Du Cane Road, London W12 0NN, UK

*β*IOS
SCIENTIFIC
PUBLISHERS

© BIOS Scientific Publishers Limited, 1999

First published 1999

A CIP catalogue record for this book is available from the British Library.

ISBN 1-85996-180-0

BIOS Scientific Publishers Ltd
9 Newtec Place, Magdalen Road, Oxford OX4 1RE, UK
Tel. +44 (0)1865 726286. Fax +44 (0)1865 246823
World Wide Web home page: http://www.bios.co.uk/

Published in the United States of America, its dependent territories and Canada by Springer-Verlag New York Inc., 175 Fifth Avenue, New York, NY 10010-7858, in association with BIOS Scientific Publishers Ltd.

Published in Hong Kong, Taiwan, Singapore, Thailand, Cambodia, Korea, The Philippines, Indonesia, The People's Republic of China, Brunei, Laos, Malaysia, Macau and Vietnam by Springer-Verlag Singapore Pte. Ltd, 1 Tannery Road, Singapore 347719, in association with BIOS Scientific Publishers Ltd.

Production Editor: Jonathan Gunning.
Typeset by Marksbury Multimedia Ltd, Midsomer Norton, Bath, UK.
Printed by The Cromwell Press, Trowbridge, UK.

8   Targeting of gene delivery systems   **125**
   L.-A. Martin
   Introduction   125
   Ex vivo gene transfer   125
   Transduction targeting   127
   Liposome vectors   133
   Molecular conjugates   134
   Targeting at the transcriptional level   135
   Conclusion   138
   Further reading   139

9   Gene therapy in the clinic: human trials of gene therapy   **141**
   T. Valere
   Introduction   141
   Gene therapy around the globe   141
   Vectors and routes of administration   143
   Which genes for which diseases?   143
   Results: the global picture   145
   Conclusion   152
   References   152
   Further reading   154

10   Ethical issues in gene therapy   **155**
   N.C. Nevin
   Introduction   155
   Supervision of gene therapy research   156
   Consent to research   157
   Confidentiality   158
   Medical surveillance of gene therapy patients   159
   Germline gene therapy   159
   In utero gene therapy   160
   Conclusion   161
   References   162

11   Prospects for gene therapy   **163**
   N.R. Lemoine
   Introduction   163
   Mechanisms of gene transfer and expression   163
   Development of models of disease and therapy   164
   Cycles of clinical development – beyond the phase I trial   164
   References   165

Contributors

Drew, J., Phogen Laboratories, Marie Curie Research Institute, Trevereux Hill, Limpsfield Chart, Oxted, Surrey RH8 0TL, UK

Fidler, S., Department of GUM/HIV, Imperial College School of Medicine, St Mary's Hospital, Praed Street, London W2 1NY, UK

Lemoine, N.R., ICRF Molecular Oncology Unit, Imperial College School of Medicine, Hammersmith Campus, Du Cane Road, London W12 0NN, UK

Martin, L.-A., Institute of Cancer Research, Department of Academic Biochemistry, Wallace Wing, Royal Marsden Hospital, Fulham Road, London SW3 6JJ, UK

Miller, A.D., Imperial College Genetic Therapies Centre, Department of Chemistry, Imperial College of Science, Technology and Medicine, South Kensington, London SW7 2AY, UK

Murphy, S., Guggenheim 1836, Mayo Foundation, 200 First Street SW, Rochester, MN 55905, USA

Nevin, N.C., Department of Medical Genetics, Floor A, Belfast City Hospital Trust, Lisburn Road, Belfast BT9 7AB, UK

Rigg, A.S., ICRF Molecular Oncology Unit, Imperial College School of Medicine, Hammersmith Campus, Du Cane Road, London W12 0NN, UK

Valere, T., Journal of Gene Medicine, 14 rue St Laurent, F-60500 Chantilly, France

Vassaux, G., ICRF Molecular Oncology Unit, Imperial College School of Medicine, Hammersmith Campus, Du Cane Road, London W12 0NN, UK

Weber, J., Division of Medicine, Imperial College School of Medicine, St Mary's Hospital, Praed Street, London W2 1NY, UK

Abbreviations

14 Dea 2	tetradecanoyl-N-(trimethylammonioacetyl) diethanolamine chloride
2C$_{12}$-L-Glu-ph-C$_2$-N$^+$	O,O′-didodecyl-N-[p-(2-trimethylammonio-ethyloxy)benzoyl]-(L)-glutamate
AAV	adeno-associated virus
Ad	adenovirus
ADA	adenosine deaminase
ADCC	antibody dependent cell-mediated cytotoxicity
AFP	α-fetoprotein
AIDS	acquired immunodeficiency syndrome
ALV	avian leukosis virus
APC	adenomatous polyposis coli
ASOR	asialoorosomucoid
BGTC	bis-guanidinium-tren-cholesterol
BHK	baby hamster kidney
BMT	bone marrow transplantation
CAR	coxsackie and adenovirus receptor
cDNA	complementary DNA
CEA	carcinoembryonic antigen
CEβA	cholesterol ester of β-alanine
CF	cystic fibrosis
CNTF	ciliary neurotrophic factor
CFTR	cystic fibrosis transmembrane conductance regulator
CHD	coronary heart disease
Chol	cholesterol
CMV	cytomegalovirus
CTAP	N^{15}-cholesteryloxycarbonyl-3,7,12-triazapenta decane-1,15-diamine
CTL	cytotoxic lymphocytes
dADO	deoxyadenosine
dATP	deoxyadenosine triphosphate
DC	dendritic cells
DCC	deleted in colorectal cancer
DC-Chol	3β-[N-(N',N'-dimethylaminoethane)carbamoyl]-cholesterol

DDAB	dimethyldioctadecylammonium bromide
di C 14 amidine	N-t-butyl-N'-tetradecyl-3-tetradecyl aminopropionamidine
DISC	disabled infectious single copy
DMD	Duchenne muscular dystrophy
DMPE-PEG$_{5000}$	dimyristoyl L-α-phosphatidylethanolamino-poly (ethylene glycol)-5000
DMRIE	1,2-dimyristyloxypropyl-3-dimethylhydroxy-ethylammonium bromide
DODAC	dioleyldimethylammonium chloride
DOGS	dioctadecylamidoglycylspermine
DOPE	dioleoyl L-α-phosphatidylethanolamine
DORI	1,2-dioleoyloxypropyl-3-dimethylhydroxyethyl-ammonium bromide
DORIE	1,2-dioleyloxypropyl-3-dimethylhydroxyethyl-ammonium bromide
DOSPA	2,3-dioleyloxy-N-[2-(sperminecarboxamido) ethyl]-N,N-dimethyl-1-propanaminium trifluoroacetate
DOSPER	1,3-dioleoyloxy-2-(6-carboxyspermyl)propylamide
DOTAP	1,2-dioleoyloxy-3-(trimethylammonio)propane
DOTIM	1-[2-(oleoyloxy)ethyl]-2-oleyl-3-(2-hydroxyethyl)-imidazolinium chloride
DOTMA	N-[1-(2,3-dioleyloxy)propyl]-N-N-N-trimethyl ammonia chloride
DPDPB	1,4-di[3',2'-pyridyldithio(propionamido)butane]
DRV	dehydration–rehydration
EBV	Epstein–Barr virus
EDMPC	1,2-dimyristoyl-sn-glycero-3-ethylphospho-choline, chloride salt
EGF	epidermal growth factor
EGFR	epidermal growth factor receptor
egg PC	egg phosphatidyl choline
EMEA	European Medicines Control Agency
eNOS	endothelial nitric oxide synthase
FAP	familial adenomatous polyposis
FDA	Food and Drug Administration
FGFR	fibroblast growth factor receptor
FH	familial hypercholesterolemia
FIV	feline immunodeficiency virus
GAP-DLRIE	($\pm$)-N-(3-aminopropyl)-N,N-dimethyl-2,3-bis (dodecyloxy)-1-propanaminium bromide
GCV	ganciclovir
GPAT	genetic prodrug activation therapy

GRP	gastrin releasing peptide
GTAC	Gene Therapy Advisory Committee
GVHD	graft-versus-host disease
HBV	hepatitis B virus
HCV	hepatitis C virus
HDL	high density lipoprotein
HIV	human immunodeficiency virus
HMG-1	high mobility group 1 protein
HNPCC	hereditary non-polyposis colorectal cancer
HSC	hematopoietic stem cell
HSV	herpes simplex virus
HSVTK	herpes simplex virus thymidine kinase
HTLV-1	human T-cell leukemia virus 1
HUVEC	human umbilical vein endothelial cells
HV	herpes virus
HVJ	hemagglutinating virus of Japan (Sendai virus)
i.m.	intramuscular
i.p.	intraperitoneal
i.v.	intravenous
IDDM	insulin-dependent diabetes mellitus
IE	immediate early
ITR	inverted terminal repeat
LCR	locus control regions
LDL	low density lipoprotein
LDLR	low density lipoprotein receptor
LOH	loss of heterozygosity analysis
LPD	lipid–protamine–DNA
L-PE	lysinyl phosphatidylethanolamine
LTR	long terminal repeat
Lys-Pam$_2$-GroPEtn	lysinyl-dipalmitoyl-(L)-α-phosphatidylethanol-amine
MCA	Medicines Control Agency
MDR	multidrug resistance
MHC	major histocompatability complex
MLP	major late promoter
MLR	major late region
MMLV	amphotropic murine leukemia virus
MMV	mouse minute virus
MoMuLV	moloney murine leukemia virus
MRI	magnetic resonance imaging
mRNA	messenger RNA
MuLV	murine leukemia virus
MVA	modified vaccinia virus Ankara

Nae	N-(2-aminoethyl)glycine
NIDDM	noninsulin-dependent diabetes mellitus
NIH	National Institutes of Health
NK	natural killer cells
NLS	nuclear localization
Npe	N-(2-phenylethyl)glycine
p(DMAEMA)	poly(2-(dimethylamino)ethyl methacrylate)
PAMAM	polyamidoamine
PAV	pseudoadenovirus
PBL	peripheral blood lymphocytes
PCR	polymerase chain reaction
PEG	polyethylene glycol
PEI	polyethylenimine
PET	positron emission tomography
PEVP	poly(N-ethyl-4-vinylpyridinium bromide)
PF	platelet factor
pfu	plaque forming unit
pLL	poly-L-lysine
RAC	Recombinant DNA Advisory Committee
Rb	retinoblastoma
RCA	replication competent adenovirus
RER	replicative errors
RES	reticuloendothelial system
REV	reverse phase evaporation
RGD	arg-gly-asp
rIL-2	recombinant human interleukine 2
RIP	rat insulin gene promoter
RRE	*rev*-reactive element
RSV	Rous sarcoma virus
RT	reverse transcriptase
RTI	reverse transcriptase inhibitor
RV	retrovirus
RVR	replication competent retrovirus
SAHH	S-adenosylhomocysteine hydrolase
SAINT	synthetic amphiphile INTeraction
SCID	severe combined immunodeficiency
SU	surface unit
TB	tuberculosis
TIL	tumor infiltrating lymphocyte
TIMP	tissue inhibitors of matrix metalloproteinases
TK	thymidine kinase
TMAG	N-(α-trimethylammonioacetyl)-didodecyl-D-glutamate chloride

TP	terminal protein
TPL	tripartite leader
TRE	transcriptional regulatory element
TSG	tumor suppressor gene
VCAM-1	vascular cell adhesion molecule-1
VEGF	vascular endothelial growth factor
VLA4	very late antigen 4
VSV	vesicular stomatitis virus

Preface

For a field that is only just 10 years old, gene therapy has already come a long way. The early clinical trials have taught us to be realistic in our expectations but we have reason to be optimistic about the future. We have witnessed spectacular advances in genomics and molecular cell biology, producing a revolution in the way in which clinicians and scientists think about disease – our new knowledge gives us the power to exploit genes themselves for therapy. With the completion of the Human Genome Project we are armed with every human gene and their control sequences for the design of tomorrow's gene therapeutics. Genetic manipulation of pluripotent bone marrow stem cells allows cell-based gene transfer to a variety of target tissues and we can anticipate developments in delivery technology using novel viral vectors and synthetic systems which will enable gene transfer with the necessary efficiency and selectivity. There are still major challenges to overcome but imaginative approaches abound along with the enthusiasm and determination to make disease cure by gene therapy a clinical reality.

N.R. Lemoine

Chapter 1

What is gene therapy?

Jeff Drew and Lesley-Ann Martin

1.1 Introduction

In 1952 Hershey and Chase made the groundbreaking discovery that DNA was the genetic material [1]. Within the next decade, the structure of the double helix was discerned [2] followed by the elucidation of the genetic code [3] establishing the new field of molecular biology and a chance to dream of its potential. Once Maniatis had cloned and manipulated the globin genes, the dream became a reality – could we use this information to treat hereditary diseases? From these fledgling findings the field of gene therapy was born.

At the most basic level, gene therapy can be described as the intracellular delivery of genetic material to generate a therapeutic effect by correcting an existing abnormality or providing cells with a new function. Initially, inherited genetic disorders were the main focus but now a wide range of diseases, including cancer, vascular disease, neurodegenerative disorders and other acquired diseases are being considered as targets.

The ultimate goal of gene therapy is the amelioration of disease upon a single administration of an appropriate therapeutic gene.

The genetic materials considered for use are intended to replace a defective or missing gene, augment the functions of the genes present, instill a specified sensitivity to a normally inert prodrug or to interfere with the life cycle of infectious diseases.

1.2 Germline gene therapy

Germline gene therapy is currently considered to be ethically unacceptable, even though it has the potential to eradicate many hereditary disorders. The technology is relatively simple and requires no targeting as genetic abnormalities can be corrected by direct manipulation. With increasingly sophisticated techniques being developed in the field of transgenic animals, it is now possible, using homologous recombination, to replace old genes for new. The results of the human genome project and establishment of gene function will open up new possibilities for genetic intervention that could be passed down through generations. It is this permanency that frightens many ethicists, and so to date all gene therapy applications have considered only somatic gene manipulation (see below).

Until such time as gene therapy for many illnesses is commonplace, with all the hurdles to routine use of gene therapy overcome and side-effects identified and dealt with, germline gene therapy will not be attempted. Ultimately, however, the question of germline gene therapy will have to be re-evaluated and we may see many of these crippling inherited diseases disappear from the gene pool. This type of scenario serves to underline the untapped potential of gene therapy. For the first time we are at a point where it is within our grasp to rid an individual and their progeny of the misery of an inherited disorder.

1.3 Somatic gene therapy

Somatic gene therapy involves the insertion of genes into diploid cells of an individual where the genetic material is not passed onto its progeny. The transfer is currently achieved by use of lipids and virus-mediated systems. To date, there are three divisions of somatic gene therapy (*Figure 1.1*).

1.3.1 Ex vivo *delivery*

In this system genetic material is delivered after explantation, cultivation and manipulation *in vitro*, followed by subsequent re-implantation. The *ex vivo* route is attractive mainly due to its lack of complicating immunological problems and enhanced efficiency of vector delivery *in vitro*. Cells can be manipulated to incorporate DNA-encoding proteins dysfunctional or lacking in the host. An example of this is the transduction of T lymphocytes to express adenosine deaminase (ADA), the enzyme known to be defective in the disorder severe combined immunodeficiency (SCID) ADA. Alternatively, in the case of familial hypercholesterolemia, secretion of protein (e.g. ApoAI/ApoE) could provide benefit by lowering serum cholesterol.

This would provide benefit as protein would be expressed at therapeutic levels for extended periods reducing the need for readministration of purified protein or drugs to control the effects of the disease. This would contribute significantly to reducing healthcare costs and improve the quality of life of many sufferers. There are, however, major constraints on the use of *ex vivo* therapy in that only some disease states are amenable to the approach. In addition, at present only a small percentage of re-implanted cells remain viable.

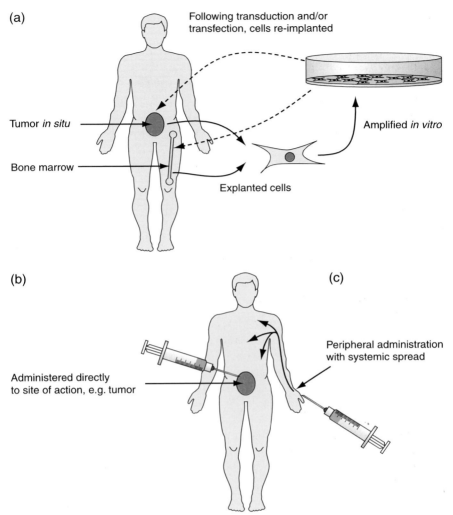

Figure 1.1. Diagrammatic representation of the different ways of applying gene therapy vectors. (a) *Ex vivo* gene therapy. (b) *In situ* gene therapy. (c) *In vivo* gene therapy.

1.3.2 In situ *delivery*

Administering the material directly to the desired tissue is currently the major area of clinical interest, as many of the current delivery systems lack effective targeting. One disease that has shown some success with this strategy has been cystic fibrosis (CF). This disorder results from mutations in the cystic fibrosis transmembrane conductance regulator (CFTR). As a consequence, mucociliary action is impeded leading to bacterial colonization of the airways. The CFTR gene has been delivered using lipid and adenoviral vectors to defined sites in the respiratory tract. Initial results demonstrated a 20–30% correction in CFTR-mediated ion transport at these sites.

This form of delivery is also used in the treatment of cancer. A study in a rodent model used direct administration of a retroviral vector expressing the suicide gene herpes simplex virus thymidine kinase (HSVTK) directly into small intracerebral gliomas. Upon treatment with the prodrug ganciclovir, 75% of the tumors showed regression.

However, low efficiency of transduction is a continuing problem. This is a particularly important consideration in cancer therapy, as a single malignant cell remaining would re-establish the tumor.

1.3.3 In vivo *delivery*

Administration of material systemically is probably the least advanced strategy at present, but potentially the most useful. The main reason for this is insufficient targeting of the vectors to the correct tissue sites. The liver also rapidly clears adenoviral vectors together with lipid/DNA complexes after administration. Current retroviral vectors activate complement resulting in their neutralization, and all of the viral vectors used to date stimulate strong immune responses upon repeated dosing. However, improvements in targeting and vector development are addressing these problems.

All gene therapy applications depend on the fact that genetic material needs to be delivered across the cell membrane and ultimately to the cell nucleus.

After enormous effort the major obstacle to harnessing the potential of gene therapy remains inefficient intracellular delivery. Significant work is underway to develop efficient gene delivery vehicles using modified virus particles or chemically synthesized entities. More recently, a possible aid to the low efficiency of gene transfer has been the identification of the unusual trafficking properties of one of the herpes simplex virus tegument proteins, VP22. This protein has the capacity to spread (either alone or as a fusion peptide) from the cells in which it is expressed, to the nuclei of surrounding cells.

1.4 Viral systems

Many viral vector systems now exist for use in gene delivery studies. The most widely studied include retrovirus, adenovirus (types 2 and 5), adeno-associated virus and herpesvirus. Common to all viral vectors is the fact that their genomes have been modified, deleting areas which render them replication incompetent. This has the effect of limiting the virus particle to only a single infectious cycle, so improving the safety of their use.

1.4.1 Retroviral vectors

Retroviruses have been extensively used for gene therapy. With a few exceptions (notably human immunodeficiency virus [HIV]) they are relatively nonpathogenic. Retroviral vectors have the ability of infecting a wide variety of cell types and have the potential of integration into host genomes. Murine leukemia virus (MuLV) has been the most widely used, with several systems developed for generation of infectious replication-incompetent particles. In essence, all of the viral genes have been removed, creating approximately 8 kb of space for transgene incorporation. Integration of the genetic material can only be accomplished in dividing cells using MuLV vectors. However, newer vector systems incorporating HIV sequences are being developed with the ability to integrate into nondividing cells. Human spumaviruses are also being examined as potential gene delivery vehicles. These are nonpathogenic human retro-viruses that have the ability to integrate into quiescent cells.

The problems with current retroviral systems include inability to produce high titers; immunogenic problems and complement inactivation in the case of MuLV. In the latter, systems are being developed to overcome this hurdle. Active research is presently being performed to address the issues that hamper retroviral vector production and use. Retroviruses are the most widely used delivery vehicle in clinical trials to date accounting for approximately 40% of studies.

1.4.2 Adenoviral vectors

The most common DNA virus system used for gene therapy applications is adenovirus (types 2 and 5). The serotypes used have not been implicated in serious illness. As an added safety measure several essential genes have been deleted so that viral replication can only occur under controlled conditions. Adenovirus-delivered genes are episomally maintained and lost due to genetic instability. Hence repeated doses are necessary to maintain expression of the transgene. Most adenoviral vector systems have a transgene capacity of approximately 7 kb. However, more recently systems have been developed in which essentially all of the viral genome has been deleted resulting in sufficient space for 35 kb of transgene sequence – the so called 'gutless' or 'pseudo' adenoviruses.

The advantages of the adenoviral system are the ability to produce high numbers of purified particles, their ability to infect a wide range of tissues and, more importantly, their tropism for the lung. This has made them particularly attractive to CF gene therapists. However, the particles are highly immunogenic which limits the number of repeated applications. Cancer gene therapists have used this to their advantage, where enhanced cell death due to immune recognition is desirable.

1.4.3 Adeno-associated viral vectors

Another DNA virus-based vector system utilizes adeno-associated virus (AAV). AAV has no known pathogenic effects and is widespread – reports suggest that approximately 80% of the population possess antibodies to the virus. A great deal of attention has been focused on this system in recent years. The wild type virus has a wide tissue tropism and its genome preferentially integrates at a specific site on chromosome 19 with no noticeable effects. The AAV vectors have been shown to be weak immunogens upon delivery to some tissues. Unfortunately recombinant AAV vectors have a much-reduced propensity for site-directed integration and studies are underway to boost this desired function. The major drawbacks of these vector systems include the complicated process of vector production (most systems necessitate the use of adenovirus to supply helper functions) and the limited transgene capacity of the particles ($\leqslant 4.8$ kb.). Whereas little can be done to extend the packaging constraint of the particles, research is ongoing to modify and improve recombinant virus production.

1.4.4 Herpes simplex viral vectors

One of the most recent viruses to emerge as a candidate for vector generation is herpes simplex virus (HSV). HSV based systems include the development of the so called disabled infectious single copy (DISC) virus which comprises a glycoprotein H defective mutant HSV genome. When propagated in complementing cells viral particles are generated which can infect subsequent cells, replicate their own genome but will not produce further infectious particles. Another intriguing system involves the packaging of a minimal HSV amplicon devoid of almost all HSV sequences that would have the potential to package in excess of 150 kb of transgene sequences.

   Major hurdles still exist to the widespread use of viral vectors and research is underway to address these. Some of the more important drawbacks are immunogenicity of the particles, packaging constraints, long-term maintenance of the genetic material and, in the case of retroviruses, random integration. Another major obstacle is the specific targeting of genetic material to the correct cells. Two avenues of research are now being extensively pursued namely transduction targeting, which concentrates upon manipulation of the viral vector to obtain delivery to specific cells, and transcriptional targeting in which the transgene expression is controlled by tissue-specific transcriptional elements.

   In the following chapters, the problems regarding the use of these vectors and current thoughts for their modification will be discussed.

1.5 Nonviral systems

An alternative to the viral strategies has been the application of chemically synthesized vehicles such as liposomes or the use of naked plasmid DNA. This has been of particular interest in the emerging field of DNA vaccination technology.

1.5.1 Lipid-mediated delivery

Over the past decade, extensive research has led to the development of chemically based delivery systems. For several reasons, cationic lipids are the preferred choice, not least due to their highly efficient nature at affording gene transfer but also the ease with which the DNA/lipid complex is formed.

Cationic liposomes usually comprise a formulation of positively charged lipid and a co-lipid required for stabilization of the liposome complex. A widely used formulation is N-[1-(2, 3-dioleyloyx)propyl]-N-N-N-trimethyl ammonia chloride (DOTMA) and dioleoyl phosphatidyletha-nolamine (DOPE). Interaction between negatively charged DNA and positively charged lipid results in the spontaneous formation of complexes. There are now many preparations, with different formulations affording a variety of properties, which are capable of delivering genetic material to cells. However, once again, problems exist regarding their use. Most notably, gene transfer *in vivo* occurs at low efficiency, although some success has been achieved in delivering material to lung and liver.

1.5.2 Naked DNA

Investigations have revealed that naked plasmid DNA can enter cells and express their genetic material, although the efficiency of uptake is poor. Plasmids ranging in size between 2 and 19 kb have been successfully delivered but the mechanism by which this occurs is undetermined. Tissues exhibiting transgene expression following plasmid DNA injection include skin, thymus, cardiac muscle, and especially striated (skeletal) muscle. Long-term expression has been observed in murine striated muscle following injection, for more than 19 months. Effectiveness of gene transfer into muscle by intra-muscular DNA injection is affected by a variety of factors. Single injection yields transgene expression in less than 1% of the total myofibers of the muscle. However, this can be improved by multiple injections. Similarly, pre-treatment of the muscle with sucrose or by degenerating the muscle prior to plasmid injection improves the overall uptake and expression of the transgene. Age and species of test subjects can also influence the efficiency of uptake. Primate muscle has a lower efficiency of DNA uptake compared to murine muscle, and young mice demonstrate improved gene transfer efficiencies. This could be due in part to variations in the muscle composition. Striated muscle is currently under investigation for use as a platform for expression of secreted therapeutic polypeptides (e.g. in hyperlipidemia) following DNA injection.

The greatest advantage of intra-muscular DNA injection is its relative simplicity. However, lack of integration into the host genome, limited tissue types to which DNA can be delivered and a low efficiency of uptake of the genes following injection are just some of the disadvantages. For some applications, such as DNA vaccination, low efficiency of DNA uptake is not a problem because low level antigen expression and presentation is sufficient to stimulate the required immune responses.

An alternative to injection of naked DNA is the use of biolistic or 'gene gun' methods of administration, a modification of a technique originally developed for gene transfer to plant cells. Gold or tungsten particles (1–3 μm diameter) are coated with plasmid DNA and accelerated to high speeds by a variety of means including electronic and helium pressure discharge, enabling the coated particles to penetrate the target tissues. The technique has been used for genetic transfer to a variety of cell types, including muscle, liver and epidermis. Some *in vivo* investigations have been performed targeting surgically exposed tissue. The drawbacks of the technique include transient gene expression and cellular damage.

1.5.3 VP22

As mentioned earlier, the herpes simplex virus I (HSV-I) tegument protein VP22 has the exciting property of spreading from the cytoplasm of the cells in which it is expressed to the nuclei of surrounding cells. The mechanism by which this function occurs is currently unknown. However, this propensity remains when fusion molecules are generated between VP22 and a cargo partner, including reporter genes such as the *aequoria victoria* green fluorescent protein, and genes currently under examination as therapeutic agents such as p53. As such it heralds the prospect of improving the numbers of cells receiving therapeutic molecules, even under conditions whereby the transfer of genetic material is quite inefficient.

Despite the difficulties with the current delivery systems many diseases have reached the stage of clinical trial using these vehicles. It will be interesting to see how these fare in a clinical setting. This will give us an insight into the factors that need to be addressed to improve these vectors.

1.6 Candidate diseases for application of gene therapy

Many disease states arise from chromosomal abnormalities, which are inherited or acquired and can be either monogenic or polygenic disorders. Almost all such illnesses are notoriously difficult to treat by conventional therapies, in most cases only supportive treatment being available. A brief list (by no means exhaustive) of some of the diseases and conditions for which clinical trials have begun is given in *Table 1.1*. The total number of conditions currently under investigation for gene therapy is extensive and only a small proportion of these have reached clinical trials. It is likely to take several more years of research before many of them reach initial therapy stages. For example, a great deal of effort is being expended on therapies for Duchenne muscular dystrophy (DMD). DMD is an X-linked recessive disorder with an incidence approaching 1 in 3500 births (up to one third due

Table 1.1. Some of the diseases and conditions for which gene therapy clinical trials are underway

Monogenic disorders	Cancers
Cystic fibrosis	Gynecological tumors
ADA–SCID	Nervous system tumors
Canavan disease	Gastrointestinal tumors
Familial hypercholesterolemia	Skin tumors
Gaucher's disease	Genito-urinary tumors
Hemophilia B	Lung tumors
X-linked SCID	Hematological malignancies
Chronic granulomatous disease	Sarcomas
	Germ cell cancers

to spontaneous mutation). The disorder renders sufferers extremely dependent upon supportive and prophylactic care but is ultimately lethal by the third decade. The gene responsible for DMD has been located at position 21q of the X chromosome. It encodes the dystrophin molecule, which has a large mRNA of 14 kb which in itself poses problems, considering the packaging constraints of many vectors as mentioned earlier. Another limitation of gene therapy for DMD is the fact that skeletal muscle accounts for approximately 65% of the body's tissue mass, therefore targeting sufficient cells for therapy will be a considerable undertaking.

1.7 Summary

Gene therapy offers the exciting potential of a new therapeutic avenue for currently untreatable diseases. Such therapeutic approaches will correct genetically derived defects 'at source', with little or no side-effects. However, before any such therapies can become a clinical reality, several substantial hurdles need to be overcome, new techniques and technologies developed and novel vector systems elucidated. Already, huge progress has been made towards achieving the first gene medicines. Contributions to the success of the field will come from a wide range of sources and disciplines. Indeed the largest project undertaken in biological research, the Human Genome Project, will offer a vast amount to gene therapy, as it identifies genes and their controlling elements, which can then be adopted for application to disease.

References

1. Hershey, A.D. and Chase, M. (1952) Independent functions of viral proteins and nucleic acid in growth of bacteriophage. *J. Gen. Physiol.*, **36**, 39–56.
2. Watson, J.D. and Crick, F.H.C. (1953) Molecular structure of nucleic acids: a structure for deoxyribose nucleic acid. *Nature,* **171**, 737–738.
3. Crick, F.H.C., Barnett, L., Brenner, S. and Watts-Tobin, R.J. (1961) General nature of the genetic code for proteins. *Nature*, **192**, 1227–1232.

Further reading

Anderson, W.F. (1998) Human gene therapy. *Nature*, **392**(Supp.), pp 25–30.

Blau, H. and Khavari, P. (1997) Gene therapy: progress, problems, prospects. *Nat. Med.*, **3,** 612–613.

Phelan, A. *et al.* (1998) Intracellular delivery of functional p53 by the herpesvirus protein VP22. *Nat. Biotechnol.*, **16,** 440–443.

Chapter 2

Why gene therapy?

Lesley-Ann Martin and Jeff Drew

2.1 Introduction

Despite advances in clinical practice, many diseases still have no cure and require expensive treatments to prolong the lifespan of patients. With life expectancy in the first world increasing due to improved standards of living, governments are faced with spiralling healthcare costs. The idea that therapeutic gene transfer may permanently cure some diseases has sparked great interest in the field of gene therapy.

Genes are the carriers of human heredity. Within our genes are instructions for the synthesis and control of a plethora of proteins, which function to maintain the homeostasis of each cell. Once a human offspring begins an independent existence, the genetic blueprint received from its parents determines many aspects of the way the course of its life will run and what illnesses will afflict it.

In recent years our understanding of genetic inheritance has made us aware of a range of genetic abnormalities arising from defects in our genetic map. Knowledge of inherited disorders such as hemophilia have been widely accepted for decades. However, more recently the genetic components of many other common diseases have been established such as cancer, heart disease and diabetes. Many of these abnormalities require environmental stimulation to initiate them, but without the genetic predisposition the disease might never develop.

Gene therapy is defined as the introduction of genetic material into cells to bring about a therapeutic effect. Examples of the applications of gene therapy include single gene disorders such as cystic fibrosis (CF) and muscular dystrophy, to replace or augment the function of the mutated gene, and cancer, to introduce a suicide gene that sensitizes tumor cells to a prodrug.

2.2 What makes a disease suitable for gene therapy?

Over 4000 diseases result from single genetic disorders. Many result from mutations in the nucleic acid sequence resulting in the synthesis of defective proteins that are unable to perform their desired function. In some cases, if the protein is sufficiently abnormal, it may be recognized as nonself and activate an immune response. It is postulated that some autoimmune diseases may occur in this way. Alternatively, the protein may have a regulatory role such as those encoded by tumor suppressor genes. Mutations in p53 are present in over 50% of cancers. p53 plays a pivotal role in arresting cell growth in response to DNA damage so that repair or apoptosis can occur.

At present, the suitability of a disease to be treated by gene therapy relies on several prerequisites. The disease must be life-threatening, thus making its treatment ethically acceptable to the risk of side-effects. The effects of the disease must be potentially reversible by the treatment. The gene must be fully characterized and the delivery to the affected site feasible. This can involve *ex vivo* transfection or transduction of cells removed from the patient, which are then returned after manipulation. This has been the approach for the treatment of diseases such as adenosine deaminase deficiency and hemophilia. Also necessary are short-term surrogate end-points to demonstrate the physiological benefit of the inserted gene. For example, in the case of CF the electrical conductance change over the nasal epithelium can be monitored.

At present only a small percentage of monogenic disorders have been investigated for gene therapy. *Table 2.1* shows a list of the most common disorders and their estimated prevalence. Each of these diseases would require lifelong gene therapy with the current vector systems. However, with further improvements in the delivery vehicles, it may be possible to maintain control with infrequent administrations, once or twice a year. The benefit of successful gene therapy for most of these conditions could be life saving, and would reduce the cost of managing these diseases with current medical practices. Consequently the cost of gene therapy, even if it remains at thousands of dollars per patient per annum, could be acceptable in comparison to the alternative.

Table 2.1. Estimated prevalence of the major monogenetic disorders

Disease	Europe	USA	Japan
Familial hypercholesterolemia	500 000	470 000	151 000
Polycystic kidney disease	166 000	156 000	50 500
Huntington's chorea	91 000	86 000	28 000
Cystic fibrosis	48 000	45 000	15 000
Hemophilia	25 000	23 000	7 500
Phenylketonuria	24 000	23 000	7 500
Duchenne muscular dystrophy	11 000	10 000	3 300

2.3 Cystic fibrosis

CF is an autosomal recessive disorder, resulting in abnormal electrolyte transport across the surface of epithelial cells. Almost 1 in 20 people are heterozygous for the disease and 1 in 2000 have CF at birth. The disease causes respiratory failure as a result of bacteria colonizing the airways. Mutations in the gene encoding the cystic fibrosis transmembrane conductance regulator (CFTR) have been pinpointed as the cause of the disease. CFTR is a cAMP-mediated chloride channel. Mutations in the protein prevent proper channel opening, inhibiting the movement of water across the epithelial surface. As a result the mucus within the airways becomes sticky, suppressing mucocilliary action. The disorder also causes blockages in the intestine and sterility. Most sufferers do not survive beyond their mid-thirties. Present treatments involve administering antibiotics and enzyme supplements. As a lethal monogeneic disorder, CF has attracted a lot of attention. Several trials have been carried out delivering the CFTR gene in lipid conjugates or encoded in adenoviruses. These initial studies used electrical conductance across the nasal membrane as a measure of successful transgene delivery and expression. The results from the adenoviral CFTR phase I study showed a 30% improvement in chloride (Cl^-) secretion over a 2-week period without any side-effects in two of the three patients. However, the third showed signs of an inflammatory response to the adenovirus vector. In the lipid–CFTR trial, patients showed a 20% improvement in Cl^- movement and no toxic side-effects were noted. Further studies are eagerly awaited.

2.4 Adenosine deaminase deficiency

Adenosine deaminase (ADA) deficiency is a rare autosomal recessive disorder that results in severe combined immunodeficiency. This disease was the first to be treated in a phase I clinical study in 1990. The disease is caused by defective expression of the enzyme ADA that plays a major role in the purine salvage pathway. Absence of the enzyme results in elevated levels of dATP blocking T-lymphocyte differentiation in the thymus. Current therapies include bone marrow transplants, which provide the most successful cure or supplements of PEG-ADA. The phase I gene therapy trial removed peripheral T lymphocytes from the patients. These were then transduced *ex vivo* with a retrovirus expressing ADA. Both patients showed a rapid increase in functional T cells, cell-mediated immunity and humoral activity. However, during the trial the patients carried on receiving PEG-ADA as requested by the ethics committee. Over the past years the patients have received several gene therapy treatments and remain responsive. The level of PEG-ADA given has not been increased from its initial dose. More recently scientists have been attempting to transduce bone marrow progenitor cells – the ultimate target.

2.5 Familial hypercholesterolemia

Familial hypercholesterolemia (FH) is a frequently occurring autosomal dominant disorder occurring in 1:500 births. The condition often results in premature development of atherosclerosis, a condition in which the arteries become thickened and eventually blocked. If this occurs in the major coronary arteries, the patient can suffer angina and acute myocardial infarction. At present, coronary heart disease is treated with bypass operations, which involve the grafting of new arteries, or by percutaneous transluminal coronary angioplasty that involves the insertion of a tiny balloon into the blocked artery which is then inflated to remove the obstruction. The latter technique is favored because it is cheaper, requires no general anesthetic and is easier. The major drawback of both procedures is restenosis, in which the arteries become reoccluded. Unfortunately this is known to occur in 30–50% of patients. Another alternative is to give lipid-lowering drugs, but some patients become refractory to these.

Gene therapy therefore, offers an exciting alternative. It is recognized that FH results from the absence of low density lipoprotein (LDL) receptors, and as a consequence the level of LDL remains elevated in the blood while high density lipoprotein (HDL) is suppressed. A phase I clinical trial is currently under way in which hepatocytes isolated from the patients' liver are transduced *ex vivo* with a retrovirus expressing LDL receptors. The modified cells are then reinfused into the inferior mesenteric vein. To date, one patient on the study has shown a marked decrease in serum cholesterol levels and an improvement in LDL:HLD ratio over the past 2 years.

2.6 Cancer – could gene therapy provide a cure?

Cancer is one of the leading causes of mortality in the developed world (*Table 2.2*). Despite advances in surgical, radiotherapy and chemotherapy techniques, the prognosis for many cancers remains poor. The major hurdle with current treatments is the inability to differentiate between normal and tumor tissue. Local therapies such as surgery and radio-therapy are effective if the tumor is retained within the treatment area, which occurs in approximately 30% of cases. However, in the majority some form of chemotherapy is required. Many systemic drugs are available but they cure only small proportions of patients. Although treatments for several cancers (childhood leukemia, Hodgkin's disease, non-Hodgkin's lymphoma, choriocarcinoma and germ cell tumors) have been successful,

Table 2.2. Estimated annual occurrence of the major cancers in Europe

Tumor type	Annual occurrence
Lung cancer	270 000
Breast cancer	250 000
Colorectal cancer	180 000
Malignant lymphoma, myeloma	45 000
Leukemias	33 000
Ovarian	30 000

the majority of common tumors (breast, lung and colon) remain difficult to cure, despite the development of new systemic drugs, novel drug combinations, increased dosing regimes and addition of cytokines.

As a consequence of the surge of information regarding the molecular basis of cancer, gene therapy provides a tempting alternative. The major hurdle that cancer gene therapists face is how to target the therapeutic gene to every tumor cell. This has proved to be quite problematic. Unlike CF in which only 5–10% of cells need to express the CFTR gene to bring about a therapeutic benefit, in the case of cancer, any remaining cells that are untreated have the potential to metastasize. At present there are five gene therapy strategies for the treatment of cancer (described in greater detail in later chapters).

2.6.1 Augmentation

Replacement or augmentation takes the line of classical gene therapy. Tumor suppressor genes such as p53 and retinoblastoma are mutated in a number of tumors. Augmentation attempts to replace the defective gene with wild type. Several studies have been reported using adenovirus and retrovirus delivery systems. Phase I clinical trials using augmentation now account for 7% of the total cancer trials to date.

2.6.2 Antisense oligonucleotides

A second method involves the use of antisense oligonucleotides specifically designed to target cellular transcription and translation. Such oligonucleotides have been used to target mutant p53 in human lung carcinomas and the BCR-ABL translocation that gives rise to chronic myeloid leukemia. More recently they have been used to target the proto-oncogene cERBB2, found to be upregulated in breast and pancreatic tumors. Mutant forms of the RAS oncogene (known to present in 75% of pancreatic tumors) have also been targeted. The major drawback of augmentation and antisense technologies is their inability to target every tumor cell with the delivery systems currently available. As it only takes a single tumor cell to form a metastasis the application of the technology at present is limited.

2.6.3 Suicide gene therapy

To date, the most promising strategy has been suicide gene therapy. This exploits the differences between normal and neoplastic cells to drive the selective expression of a metabolic suicide gene conferring sensitivity to a prodrug. The suicide genes used generally come from bacteria or viruses, e.g. the thymidine kinase (TK) gene from herpes simplex virus. This is able to metabolize the prodrug ganciclovir into a toxic form which results in the death of herpes-infected cells. If this gene is placed in tumor cells, these become sensitized to ganciclovir while normal cells are unaffected. This strategy has generated great excitement and is now employed in over 14% of cancer gene therapy trials. The major drawback is that not all the tumor cells can be targeted, but unlike augmentation and antisense therapies, the bystander effect in this protocol is very strong. Bystander effect relies on the ability of cells to communicate via gap junctions in the cell membrane. Consequently, tumor cells that are in the vicinity of those producing the toxic metabolite will also die. More recently an immune component which enhances tumor death has also been noted, making this an extremely powerful system.

2.6.4 Immunotherapy

The majority of cancer gene therapy trials (48%) are for immunotherapy. It has been recognized for many years that tumor cells provide a very weak immune response; first, because they are 'self', and second, by possessing means for down-regulating the immune system. Hence the potential for stimulating an immune response to recognize them as 'danger' has tempted scientists for many years. Major studies have involved *ex vivo* manipulation of tumor cells to express cytokines such as IL-2, gamma interferon and tumor necrosis factor to stimulate tumor immunity. Using irradiated tumor cells as a vaccine has also proved popular, particularly for the treatment of melanoma. More recently we have seen a surge in the use of DNA vaccines encoding known tumor antigens.

At the forefront of cancer immunotherapy at present is the use of professional antigen presenting cells, e.g. dendritic cells. These are isolated from the patient's blood and then pulsed with tumor lysates or peptides prior to their systemic delivery, in the hope that they will elicit an immune response. Akin to this, dendritic cells are now being transduced with vectors expressing known tumor antigens for use as vaccines.

2.7 Infectious diseases

Gene therapy provides a very tempting prospect for the treatment of infectious diseases such as acquired immunodeficiency syndrome (AIDS) which afflicts 14 million people world-wide. This results from the infection of patients with the human immunodeficiency virus (HIV). The virus binds to the $CD4^+$ receptor on T cells, causing a rapid loss in immunity. The disease generally presents symptoms 10 years after contraction, when patients suffer from pneumonia, tuberculosis, herpes simplex infections and cancers such as Kaposi sarcoma, followed by death. The effects of the disease are currently suppressed by administering drugs such as AZT. However, to date no treatment has been developed which has the potential to save patient lives. As no cure is available, gene therapy studies have been ongoing in this area. These have included antisense therapies to block virus replication, the use of suicide genes delivered by adeno-associated viruses which cause the HIV to self-destruct, and transduction of fibroblasts from asymptomatic patients using retroviruses encoding HIV coat proteins as a vaccine to elicit an immune response.

Advances in molecular biology have made the possibility of DNA vaccination against infectious disease feasible. These systems rely on the expression of viral or bacterial genes from expression plasmids. The advantage of this system compared to the current use of (for instance) attenuated viral vaccines, is that the cold chain necessary for viral vaccine transport is unnecessary. DNA vaccines can be lyophilized and reconstituted prior to immunization making the system more cost effective.

2.8 Ethical and regulatory considerations

It is widely recognized that the major risk of *in vivo* gene manipulation is the insertion of genetic material up- or down-stream of tumor suppressor genes or oncogenes, resulting in the development of neoplasia. As a consequence, the use of retroviruses that integrate genetic material in the host genome have caused the most concern. However, the chance of an integration event activating an oncogene is estimated at less than one in a million. Other potential risks from gene therapy include recombination of the disabled vector, resulting in the generation of an infectious virus, toxic shock as a result of viremia or physiological effects due to over-expression of the therapeutic gene.

At present it is wildly optimistic to consider any of the treatments currently under clinical investigation as cures. In recent months we have seen the results from several phase I clinical trials published. In each case some degree of successful gene transfer and even prolonged expression has been shown, but as yet all have failed to demonstrate marked clinical benefit to treated patients.

2.9 Gene therapy industry – an explosion waiting to happen?

Gene therapy, although a very exciting area of research, is high risk commercially. Many of the biotechnology companies investing in gene therapy are headed by academic team leaders exploiting their findings, or collaborations between pharmaceutical companies and academic institutions. With regard to the technology the majority of the strategies for genetic intervention have been established, with companies focusing on individual areas of application (*Table 2.3*). The major stumbling block in the development of effective gene therapy is the delivery of the therapeutic gene to its target. At present lipid systems provide the greatest safety, but poor efficacy for delivery. Viral systems, although lacking this safety, are

Table 2.3. Areas of gene therapy interest for several biotechnology concerns

Company	Technology under investigation
Applied Immune Sciences Inc.	Vector research, cell sorting and expansion, gene therapy
Introgen Therapeutics Inc.	Gene therapy, vector development p53 and K-ras gene therapies
Genetix Pharmaceuticals Inc.	Multi-drug resistance gene therapy, retroviral vector development
Institut Pasteur	Cardiovascular research
Institut Gustave Roussy	Adenoviral vectors
Transgene	Adenoviral vectors and gene therapy
Genzyme	Adenoviral vectors and lipid delivery systems for gene therapy
Chiron	Development of DNA vaccines
Genopoietic	Vector technology for gene therapy HSVTK therapies
Hybridon	Antisense technology for gene therapy
Lynx	Oligonucleotide therapies
Isis	Oligonucleotides for use as antivirals
Theragen	Retroviral vectors, adeno-associated viral vectors. Therapies for Gaucher's disease, rheumatoid arthritis, Alzheimer's, Parkinson's disease and brain tumors
Vical	DNA vaccination against infectious diseases
Phogen	Herpes simplex VP22 delivery systems
ONYX	ONYX E1B deleted adenovirus for gene therapy

superior in terms of gene delivery. As a consequence, many companies are concentrating on the development of effective delivery systems as opposed to therapeutic genes. It is clear that the first company to develop an effective delivery system will benefit financially. Among companies paying particular attention to delivery are RPR Gencell, GeneMedicine, Therexsys and Viagene.

Current research suggests that gene therapy will provide treatment for serious diseases such as cancer, and a means of preventing illness such as diabetes and hypercholesterolemia that have a genetic component. Both have market potential, particularly in the case of cancer that would be taken up irrespective of price. Current market prediction suggests revenue in excess of $12.4 million per annum in Europe. Similarly, reducing the risk of diseases such as diabetes by even 50% would secure widespread acceptance, assuming the cost could be supported and shown to reduce the pressure on health care budgets of caring for people who may otherwise develop the disease. One area which has also generated great interest is the treatment of infectious diseases, especially HIV. It has been postulated that this may be one of the first diseases treatable by gene therapy to reach the commercial market. Conservative estimates place the revenue from this between $600m in the US and $560m per annum in Europe. Paul Evers, a market research consultant, made some estimations on the value of the gene therapy market for the treatment of several diseases (*Table 2.4*). It is not surprising that many of the large

Table 2.4. Predicted European annual market value ($m) for the treatment of several target diseases by gene therapy

Disease	Market value ($m)
Lung cancer*	2700
Breast cancer*	2500
Colorectal cancer*	1800
Malignant lymphoma, myeloma*	450
Leukemias*	330
Ovarian cancer*	300
Malignant melanoma*	270
Familial hypercholesterolemia†	25.0
Polycystic kidney disease†	10.0
Cystic fibrosis†	5.2
Huntington's chorea†	5.0
Duchenne muscular dystrophy†	1.8
Hemophilia†	1.3
Phenylketonuria†	1.1

Data modified from Evers (1995).
*Cost of therapy estimated at $10 000 per patient.
†Cost of therapy estimated at $3000 per patient.

pharmaceutical companies are watching the success of the current research so avidly.

2.10 Future prospects for gene therapy

Although human gene therapy is still in its infancy, it is hard to deny that this approach has the potential to revolutionize the way in which we treat disease. The rapid advances in molecular biology and virology suggest that it is only a matter of time before we see the dream realized.

Further reading

Blau, H. and Khavari, P. (1997) Gene therapy: Progress, problems, prospects. *Nat. Med.*, **3**, 612–613.

Evers, P. (1995) *Gene Therapy: Present Status and Prospects.* Financial Times Management Report. Financial Times Pharmaceutical and Healthcare Publishing. Pearsons Professional Ltd, London.

Friedman, T. (1996) Human gene therapy – an immature genie but certainly out of the bottle. *Nat. Med.*, **2**, 144–147.

Chapter 3

Viral delivery systems for gene therapy

Stephen J. Murphy

3.1 Introduction

Although the concept of gene therapy has been established over the past decade, the establishment of effective clinical tools and techniques to facilitate an efficacious reversal of disease has proven highly problematic. The development of an effective gene delivery system to the site of therapeutic significance has proven to be the major hurdle to the advancement of gene therapies.

The advances in recombinant DNA technology have also enabled the genetic characterization of disease-inducing pathogens at the genetic level aiding the establishment of therapies to counteract their pathogenic effects. Viruses are infective agents smaller than common microorganisms, requiring living cells for multiplication. Hundreds of viruses are known, each of which infects a particular animal or plant cell. Many viruses are responsible for serious disease but others are benign intracellular parasites. Viruses were originally studied as a means of understanding disease, however elucidation of the complex life cycles of a number of viruses realized their potential as vehicles for delivery of therapeutic genetic material. Viruses have throughout evolution developed highly effective methods of entering cells, evading the host immune defence and delivering their viral message to the nucleus. Hence an immense amount of research has been directed at harnessing this viral property, to deliver a viral genome containing foreign genes using genetically modified viruses. Our present understanding of the molecular genetics of many viruses renders possible their manipulation as cloning vectors for gene transfer both in cell culture and in animals. As the major objective is long-lasting gene transfer, a deletion of the key regulatory viral genes is essential to manipulate the genetic program of the virus and to ensure that infection

21

of the target cell does not lead to cell death.) Care must also be taken in the construction of viral cloning vectors to ensure that the virus is replication-deficient and that wild-type propagation is not initiated, causing infection of the surrounding cells and tissues.

Currently gene therapy is aimed at the level of gene augmentation, where a normal gene is introduced into cells such that it can function to produce sufficient quantities of the correct gene product to compensate for the lack of expression of the mutant host gene. Recombinant viral vectors are generally constructed by insertion of a gene of interest, together with the required regulatory elements, into deleted regions of the wild type viral genome. The deleted regions generally encompass essential regulatory regions of the viral genome, which must be supplied in *trans* to enable the assembly of infectious viral particles *in vitro*. *Trans*-complementing cell lines are generally employed which have been stably transformed with the essential viral coding cassettes deleted from the recombinant vectors (*Figure 3.1*). Hence transfection of the recombinant viral vector genomes into these packaging cells enables the assembly of infectious viral particles which can only replicate in cells *trans*-complementing the deleted gene functions. Therefore in a gene therapy application the recombinant viruses can infect the targeted tissue as wild type virus and express the *trans*-gene of interest, but cannot replicate or generate infectious progeny virus.

The ultimate goal is long-term, sufficiently regulated expression of the transferred gene, achieved by a single, lifetime treatment involving a simple, noninvasive, safe and efficient gene delivery which can be incorporated into clinical practice. Viruses have the potential of meeting all these criteria, although at present no individual virus system can meet them all. The immune system currently poses the major obstacle to the efficacy of most recombinant viral vectors, being naturally primed to eliminate viral infections. As our knowledge of the immune system expands, techniques of manipulating the immune system or modifying viruses to reduce their immunogenicity are being developed. Immuno-suppressive agents are being developed aimed at attenuating the specific immune functions which target the viral vectors while not compromising the immune system to such an extent that patients are left vulnerable to opportunistic infections.

A number of viruses are being developed as gene therapy vehicles of which retroviruses (RV), adenoviruses (Ad), herpes viruses (HV) and adeno-associated viruses (AAV) are generating most interest (*Table 3.1*). The following section outlines the current state of each of these viral systems.

3.2 Retrovirus-based vectors

Retroviruses are enveloped RNA-containing eukaryotic viruses which replicate through DNA intermediates, facilitated by an RNA-directed DNA polymerase (reverse transcriptase). The reverse transcribed DNA

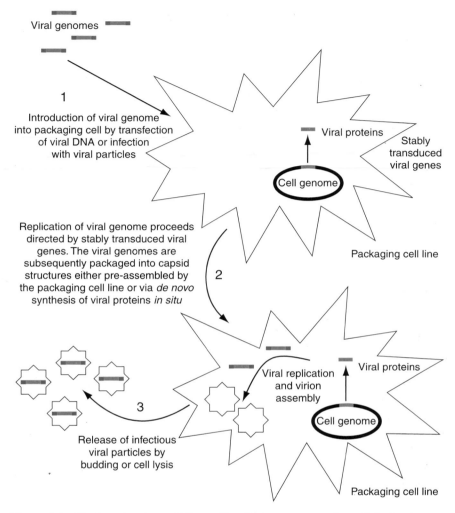

Viral genomes

1

Introduction of viral genome
into packaging cell by transfection
of viral DNA or infection
with viral particles

Viral proteins

Stably
transduced
viral genes

Cell genome

Replication of viral genome proceeds
directed by stably transduced viral
genes. The viral genomes are
subsequently packaged into capsid
structures either pre-assembled by
the packaging cell line or via *de novo*
synthesis of viral proteins *in situ*

2

Packaging cell line

Viral replication
and virion
assembly

Viral proteins

Cell genome

3

Release of infectious
viral particles by
budding or cell lysis

Packaging cell line

Figure 3.1. Viral packaging cell lines. Simplistic representation of the basic
principles of *in vitro* viral vector production in a packaging cell line. The
packaging cell line can express either one or more essential viral gene product(s),
such as with current adenovirus and herpesvirus vector technology or the majority
of the viral structural and regulatory genes as in the case of retroviral and adeno-
associated vector technology (see later sections).

genomes are subsequently integrated into the host chromosomes as
proviruses, directed by the virus-encoded integrase protein, and this DNA
is transcribed using the host machinery [1]. Retroviral genomes contain
three core genes termed *gag, pol* and *env*, (*Figure 3.2*) which are flanked by
long terminal repeat (LTR) sequences and a packaging signal which
directs the assembly of the genome into the viral particles [2]. The *gag* gene

Table 3.1. The major properties of the four most reported gene therapy viral vectors

Virus	Retrovirus	Adenovirus	Adeno-associated virus (AAV)	Herpes simplex virus (HSV)
Structure	Enveloped	Nonenveloped	Nonenveloped	Enveloped
Maximum insert capacity (kb)	9–10	7–8 helper-independent 37 helper-dependent	4.5	Large capacity Up to 150: gutless-helper-dependent
Infects quiescent cells	Type C (e.g. MLV): No Lentiviruses (e.g. HIV): Yes	Yes	Yes	Yes
Integrating	Yes	No	Yes	No
Theoretical titers	10^8–10^9	10^{11}–10^{12}	10^9–10^{10}	10^8–10^9
Genome size (kb)	~ 10	~ 36	~ 4.7 (AAV-2)	152 (HSV-1)
Particle size (nm)	Pleomorphic	60–90	18–26	~ 110

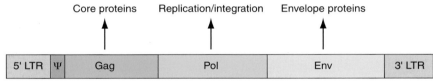

Single-stranded RNA

Figure 3.2. Retroviral genome. Schematic representation of the basic retrovirus genome, detailing the three major gag, pol and env coding units, flanked by the major regulatory long terminal repeat (LTR) elements and the packaging signal (Ψ). Also associated with other retroviral genomes are accessory genes which are not depicted in the diagram, but are generally located flanking the env gene coding unit [3].

encodes proteins which form the viral core, while the *pol* gene encodes reverse transcriptase (RT), the viral integrase (Int) and a viral protease which acts on the *gag* gene products [3]. The *env* gene encodes the glycosylated envelope proteins which determine the tropism of the virus. The LTR sequences contain the *cis*-acting sequences required to regulate viral genome replication, transcription and mediate integration into the host genome [2]. The majority of retroviruses do not cause cancers but simply infect a cell and produce a persistent infection with continual production of virus by infected cells. A number of retroviruses are, however, tumorigenic, such as the well characterized Rous sarcoma virus (RSV), which causes a connective tissue cancer in chickens. The tumorigenicity of oncogenic retroviruses can be related to accessory oncogenes, additional to the *gag, pol* and *env* genes, which convert the cell to a cancer cell, capable of indefinite growth and eventually killing the organism containing it [3]. The human immunodeficiency viruses (HIV), the causative agents of AIDS, are also retroviruses of the lentivirus subgroup, which contain several additional accessory genes [4].

Most current retroviral vectors for gene therapy purposes have been based on the well studied retroviral oncogenic subgroup Moloney murine leukemia virus (MoMuLV, [5]). The construction of retroviral-based vectors basically involves the replacement of the *gag, pol* and *env* coding units with a transgene of interest, while retaining the LTRs and packaging *cis*-acting sequences. Producer cell lines stably transformed with independent *gag/pol* and *env* expression cassettes are used to fully complement the viral polypeptides for packaging of the vector proviruses [6]. Hence by transfecting LTR-flanked retroviral cassettes into these packaging cell lines, retroviral particles bud from the host cells containing the recombinant retroviral genome capable of expressing the incorporated therapeutic gene of interest. These retroviral particles are capable of infecting cells as wild-type virus and directing the expression of the transgene of interest, but cannot replicate or generate progeny virus, preventing the conventional disease pathway of the virus.

In terms of somatic gene transfer, replication-defective retroviruses offer a number of advantages over current delivery techniques. As well as providing significantly elevated targeting efficiencies and encoding no cytotoxic or immunogenic viral antigens, they result in efficient and stable integration of the transgene into the host chromosome, enabling a permanency that facilitates long-term expression [7]. The targeting of retroviruses to specific cell populations is also being developed by engineering heterologous protein domains into the envelope glycoproteins modifying their tropism [8]. However a number of fundamental obstacles limit the effectiveness of retrovirus gene transfer, including their inability to infect post-mitotic tissues such as muscle, due to their dependence on mitosis to gain entrance to the nucleus of infected cells to facilitate integration [9]. Size limitations (9–10 kb) on the amount of DNA which can be stably incorporated into the retroviral cassettes to allow packaging into the retroviral particles, limit the application of larger genes or large regulatory elements. Additionally, the fairly low titers ($< 10^9$ infectious units per ml [iu/ml]) and the relative instability of the virions further hampers the applicability of retroviruses to gene therapy protocols. Sensitivity of murine retroviruses to inactivation by the human complement-mediated lysis pathway of the immune system has also proved problematic [10]. Current research is attempting to circumvent this by using new packaging cell lines of human origin [11] and by modification of envelope glycoproteins to produce complement resistant pseudotype retroviruses [12].

An *ex-vivo* approach of retroviral-mediated gene transfer into primary cultures of patient origin *in vitro* and autologous grafting back into the patient has produced some success [13]. While this technique avoids complement-mediated lysis it presents limitations in terms of obtaining enough tissue of patient origin to culture, which is not viable in many disease states. An alternative approach of direct injection of mitotically inactivated retroviral producer cell lines into the targeted tissue has resulted in much increased efficiencies of transduction, specifically in regenerating muscle tissue, although complications of tissue rejection need to be addressed [14].

Although a number of *in vivo* trials have resulted in stable expression for periods of up to 1 year, a major question arises about the longevity of vector gene expression in transduced somatic cells. The complication of vector promoter shut-off in stably transduced cells *in vivo* has proved problematic [15]. New vector designs containing cell-type specific promoters may be useful for providing sustained vector gene expression. The major limitation of conventional MuLV-based vectors, targeting only mitotically active cells, has limited gene delivery into targets such as neurons, hepatocytes, myocytes and hematopoietic stem cells [16]. Currently retroviral vectors derived from the lentiviruses group (such as HIV-1) which contain a number of additional accessory genes to the conventional oncogenic group retroviruses (such as MoMuLV and RSV) and involve additional nuclear localization functions, are being developed. The karyotropic property of lentiviruses provide a promising tool to facilitate retroviral mediated gene therapies to non-dividing cells [16]. However, the extremely pathogenic nature of the currently isolated lentiviral subtypes, calls for extreme caution in the application of such vectors in gene therapy protocols. Alternative lentiviruses to HIV are being investigated, such as the feline immunodeficiency virus (FIV), which have negligible/reduced pathogenicity in human hosts [17].

The risk of insertional inactivation of essential genes or activation of cellular oncogenes upon integration into the host genome remains a major concern for retroviral-mediated gene therapies. The possible contamination of retroviral stocks with replication-competent retroviruses (RVR) is also a major concern, which could be generated from packaging cell lines harboring RVRs [18]. As genetic variation occurs during retroviral replication, proviral clones need to be analyzed to determine that the provirus has the expected structure [1]. Self-inactivating retroviruses with deletions in the 3′ LTR, which are transferred to the 5′ LTR upon replication, deleting the LTR promoter and strongly inhibiting the expression of full length viral RNA offer an additional level of protection against recombinant retrovirus dissemination [18]. These vectors, however, have been reported to have high rates of recombination during transfection and cell propagation resulting in the regeneration of the U3 sequences [19]. In summary, many refinements of *in vivo* strategies are required before effective clinical trials of retrovirus-mediated gene therapy are possible.

3.3 Adenovirus-based vectors

Increasing attention has been focusing on the potential of adenoviruses as transducing viruses for use in gene therapy. Vectors based on recombinant forms of adenovirus have emerged as vehicles of choice for many applications of *in vivo* gene therapy [20]. Over the past decade an overwhelming amount of data and literature on adenovirus vectors has amassed from both preclinical and clinical studies. Adenoviruses belong to the family Adenoviridae, which is divided into two genera, *Mastadenovirus* and *Aviadenovirus* [21]. Human adenoviruses are found worldwide as 47 different serotypes, which are grouped into six sub-groups termed A–F according to their physical and pathological proper-ties [21]. Adenoviruses of subgenus C are deemed better suited as vectors for gene therapy and recombinant adenovirus vaccines, as they are not linked with serious complications in respiratory or gastroenteric infections associated with other adenovirus subgenus [20]. Additionally, adeno-viruses of subgenus C have also been well studied in virology and molecular biology, presenting a considerable amount of knowledge on their manipula-tion and applications. The members of subgenus C; Ad1, Ad2, Ad5 and Ad6 are regarded as endemic in the human population.

The adenovirus structure consists of a nonenveloped 'spiked' regular icosahedron of 60–90 nm in diameter, encompassing 20 triangular surfaces (facets) and 12 vertices [22]. The virions are composed of a protein shell, with no membrane or lipid content, surrounding a DNA-containing inner core [22]. The icosahedral capsid structure of adenovirus consists of three main proteins, termed hexon (967 amino acids), penton (571 amino acids) and fiber (582 amino acids). A total of 720 individual hexon proteins are located in each capsid, which associate as 240 trimers, termed hexomers [22]. The hexomers form the basis of the icosahedral structure, with 12 hexomers located at each facet. The penton proteins associate as pentomers, which form the basis of the 12 vertices of the icosahedrons. The fiber proteins assemble as trimers which associate with each of the 12 penton vertices, protruding from the capsid structures as the 'spike' motifs [22]. While the hexon proteins are believed to facilitate a solely structural role, the penton and fiber subunits constitute the infection machinery of the adenovirus particles. The projecting fiber proteins, which differ in length among different serotypes [23] mediate the initial attachment of the virus to cell surface receptors of which the Ad2 and Ad5 subgroups target has very recently been determined [24]. Virus is then internalized involving a secondary interaction between the penton base and the integrin family of cell surface heterodimers [25].

The adenovirus genomes consist of a single double-stranded linear DNA molecule of approximately 36 kb in length, conventionally divided into 100 map units (mu). The genome is functionally divided into two major noncontiguous overlapping regions, early and late, defined by the onset time of transcription after infection [22]. There are five distinct early regions (E1A, E1B, E2, E3 and E4) and one major late region (MLR) with five principal coding units (L1 to L5), plus several minor intermediate and/or late regions, some of which are less well characterized [22] (*Figure 3.3*).

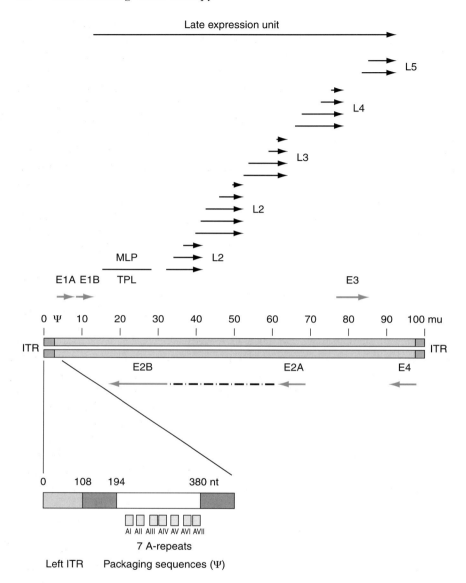

Figure 3.3. Adenovirus genome. The adenovirus genome (36 kb) is presented divided into 100 map units (mu). The four early coding units are presented (orange arrows) of which the E1 and E3 genes are transcribed from the upper rightward reading strand and the E2 and E4 genes from the lower leftward reading strand. The late coding region is presented (black arrows) and divided into the five major coding units L1 to L5, which contain the additional coding units indicated. The late region is controlled by the major late promoter (MLP) and each late transcript contains the tripartite leader sequences (TPL) shown. The genome is flanked by the *cis*-acting inverted terminal repeats (ITRs) and Ψ packaging sequences, of which the left regulatory region is shown in detail. Other minor coding units present in the complex adenovirus genome are not shown [20,21].

Both of the viral DNA strands are transcribed, with the coding units divided between the rightward and leftward reading strands according to *Figure 3.3*. Each of the adenovirus genes are transcribed by RNA polymerase II which gives rise to multiple mRNAs that are differentiated by alternative splicing. At the extremities of the viral chromosome are the inverted terminal repeats (ITRs), consisting of 100–140 bp long redundancy, which constitute the origin of replication of the adenoviral genomes. Several specific sequences located adjacent to the 5′ ITR at positions 194–385 bp (*Figure 3.3*), known as the encapsidation signals (Ψ), are essential for entry of cellular DNA into empty adenovirus virion capsids [26]. Together, the ITRs and Ψ, encompass the *cis*-elements necessary for viral DNA replication and packaging. At the 5′ termini of each DNA strand a 55 kDa terminal protein (TP) is covalently bound. An 89 kDa precursor form of this protein, pTP, functions as primers of DNA replication, and are cleaved by a virally encoded protease late in infection to the encapsidated TP forms.

Upon delivery of the adenovirus genome to the nucleus of infected cells the E1A gene (1.3–4.5 mu) is the first viral transcript to be expressed [22]. The E1A gene products act to induce expression of the other early genes, culminating in late gene expression, virion assembly and packaging. As the E1 gene products are central to these functions, first generation adenovirus vectors focused on the deletion of the E1 region, theoretically rendering the adenovirus unable to replicate in non-E1 complementing cell lines [27]. The development of the 293 human embryonic kidney cell line, which constitutively express the E1 gene products, facilitated the propagation of these replication incompetent E1-deleted adenoviruses [28].

Adenovirus vectors are attractive candidates for gene transfer due to their independence of host cell replication and low pathogenicity in man. Adenoviruses form relatively stable particles which can be readily obtained at high titers of 10^{11} to 10^{12} plaque forming units (pfu) per ml making systemic clinical administration feasible. The adenovirus genome is also relatively easy to manipulate, facilitating ease of vector generation, and does not normally undergo rearrangement at a high rate [29]. The adenovirus genome rarely integrates into the host chromosome and therefore replicates in an extra-chromosomal state, minimizing genotoxicity associated with insertional mutagenesis [22]. Questions do, however, arise concerning long-term expression from the episomally expressed genome. One major obstacle with current adenovirus vectors is the relatively tight constraints on the size of DNA which can be stably packaged into virions, of up to approximately 105% of the wild type genome [27]. This restricts the size of foreign DNA insert to about 7.8 kb [29].

The major clinical interest in adenovirus vectors stems from their broad host range and high infectivity *in vitro* and *in vivo*, exhibiting tropism for most cells in the human body, and the ability to infect quiescent as well as dividing cells [30]. Following a systemic administration of a recombinant adenovirus vector a large proportion of cells are generally found to be infected and to express the transferred gene, but to varying extents [31]. Respiratory epithelial cells are the primary target of infection, but other major sites include the eye and the gastrointestinal and urinary tracts [21,32]. Most adenoviral infections are, however, subclinical and only result in antibody formation [33]. The application of live adenovirus as oral vaccines over the past 30 years (including trials previous to license) to millions of military recruits, without any significant attributable complications, provides definitive proof for the clinical safety of human adenovirus specifically as recombinant vaccines [20]. The wide spectrum of human cells and tissues receptive to infection by adenovirus vectors and the limited pathogenicity, has made adenovirus a major candidate as a gene therapy vehicle for a large number of monogenic hereditary diseases.

The extremely efficient gene delivery with recombinant adenoviruses is not accompanied by the desired long-term gene expression. The major obstacle to current adenoviral vectors has stemmed from the significant immunogenicity they stimulate *in vivo*. Wild-type adenovirus infection results in the generation of a large number of gene products, many of which are extremely toxic and result in significant immune response. Although deleted for E1 function, first generation adenovirus vectors were discovered to result in low level leaky expression of the remaining viral genes initiated by host E1-like transcription factors [34]. Host immune responses to these foreign viral antigens have been observed to result in transient gene expression *in vivo*, with maximal expression peaking during the first week of infection followed by a rapid decline to near baseline levels within several weeks [34–36]. With few exceptions, recombinant adenovirus-mediated gene expression has not been stable and substantial pathology is observed to develop at the site of gene transfer, ultimately leading to the loss of transgene expression due to eradication of successfully transduced cells [36]. Recent reports have demonstrated that the expression of foreign genes could be observed for extended periods especially in liver and skeletal muscle [37], however, this expression is seldom detected for periods longer than 1 year.

As adenoviral genomes replicate extrachromosomally, questions arise about the long-term expression and the need for repeated administrations. Humoral immune responses of recipients against adenovirus particles form a protective barrier to repeated administrations of adenovirus vectors due to neutralizing antibodies blocking repeated entry and delivery of the vector [38,39]. Activation of T helper and B cells to viral capsid proteins leads to neutralizing antibodies that effectively block readministration of virus [38]. Other factors which contribute to the type of response elicited by the host include genotype and MHC haplotype of the recipient, target organ, dose and route of administration of the vector as well as the structure of the vector.

Current research into adenovirus vectors is focusing on strategies of circumventing host immune responses to attain long-term persistent transgene expression. Second generation adenovirus vectors are being designed which are less immunogenic to the infected host, aiming specifically at the complete ablation of expression of the immunodominant viral proteins facilitating persistent long-term expression.

The deletion of additional adenoviral regulatory genes enabled further attenuation of expression from the remaining adenoviral genes in the recombinant vectors. The construction of cell lines expressing essential gene products from the E2 [40,41] or E4 [42,43] regions, in addition to the E1 gene products, enabled further deletions of these genes. Expression of adenoviral late gene products was demonstrated to be almost completely ablated in these vectors. As well as minimizing the leaky expression of highly immunogenic adenoviral proteins, these deletions increased the cloning capacity of the vectors enabling the insertion of larger transgene cassettes. In terms of clinical safety the deletion of more than one essential regulatory element would also be advantageous in reducing the risk of the emergence of replication competent adenovirus (RCA) from genetic recombinations with adenovirus sequences in the *trans*-complementing cell line or from endemic wild type adenovirus infections.

The E3 region encodes viral proteins which play a major role *in vivo* in the persistence of adenovirus infection. As this region is nonessential to adenoviral replication and expression *in vitro*, first generation adenovirus vectors generally had this region deleted to increase the cloning capacities [27]. Second generation adenovirus vectors, however, are being engineered containing minimal E3 regions. Specifically the E3 19 kDa protein coding sequences which acts to inhibit CTL induced lysis of adenovirus infected cells by interfering with MHC class I processing [44].

Reducing the immunogenicity of the adenovirus vectors may go some way to blunt the cellular immune responses, but as the capsid structures are identical to those of wild-type adenovirus, ablation of the humoral responses is more complex. Altering external viral motifs, possibly in context with the targeting strategies, incorporating endogenous receptor ligands to the fiber motifs, may be effective at reducing the extent of neutralizing antibody production to some degree. Most strategies have, however, focused on immunosuppression. The use of immunosuppressive agents, such as cyclosporin, cyclophospamide, FK505 and deoxyspergualin have been investigated [45–48]. However, the use of these compounds may compromise the host immune system, making them susceptible to other opportune infections and have the potential of causing many severe side-effects with prolonged use [20].

Next generation adenovirus vectors are currently under a rapid phase of development involving the complete elimination of all adenoviral coding sequences, retaining only the *cis*-acting elements essential for adenoviral replication and packaging [49–53]. These so-called pseudoadenovirus

(PAV) or gutless vectors, may be rescued and propagated to high titers upon *trans*-complementation of the adenovirus genes from an adenovirus helper genome(s) co-introduced into helper permissive cell lines. These helper genomes can supply all the capsid proteins in *trans* and direct the assembly of empty adenoviral virions into which the PAV vector genomes can be packaged. PAV virions will have the same tropism and physical properties as conventional adenovirus vectors and hence high titer purified stocks could be applied to the established clinical protocols. The origins of Ad DNA replication are localized in the ITR motifs at each termini of the genome, while the packaging domain, Ψ, is located between 194 and 385 bp (0.5 and 1.1 mu) at the left end of the viral genome [22]. The sequences required in *cis* for replication and packaging of adenovirus DNA comprise less than 500 bp of the adenovirus genome [26]. Hence minimal sequence adenoviral vectors have the capacity to accommodate up to 37 kb of exogenous DNA sequence. Therefore, as well as eliminating the intrinsic toxicity or immunogenicity associated with adenovirus gene products, these minimal sequence adenovirus vectors also offer superior cloning capacities, permitting the delivery of multiple or large genes in one vector together with all the *cis*-acting elements essential to direct regulated expression. The development of minimal sequence adenovirus vectors also led to the determination of a lower size limit for adenovirus vectors for efficient packaging into the capsid particles of $\sim 75\%$ of the wild-type size [54]. Hence adenoviral vectors must lie within a 27 to 38 kb limit to be efficiently packaged, and deviation from this defined range results in significantly reduced viral titers [27,54].

The improvements in vector design associated with the PAV system, however, only serve to minimize cellular immune responses to viral protein expression and will not address the humoral or innate immune responses specific to the physical presence of the adenovirus particles. Hence the PAV vectors are susceptible to the same barriers to repeated administration as the conventional adenovirus vectors. Transient blunting or manipulation of the host immune response to the PAV vectors will thus be necessary at each administration [47].

Adenoviruses have been demonstrated to efficiently infect a great many tissue types upon systemic administration, limiting the *in vivo* targeting of gene therapy to specific cell populations. Both the fiber and penton cell surface receptors are broadly distributed over a wide tissue range; facilitating the characteristic promiscuity of adenovirus infection [55]. Alterations to the integrin binding motifs of the penton subunits have been demonstrated to redirect the specificity of penton association with alternative tissue-specific integrins, which could be of great value in gene therapy targeting of adenovirus vectors [55]. To efficiently target an adenovirus, however, similar tissue specificity would be required to be bestowed on the fiber motifs, which direct the primary receptor-mediated attachment to cells. Retargeting of the fiber ligands to alternative receptors has also been investigated involving fiber fusion proteins in which the C-terminal knob domain is replaced with alternative receptor ligands [56].

3.4 Adeno-associated virus-based vectors

Adeno-associated viruses (AAV) have recently become attractive candidates for gene transfer [57]. AAV belong to the family Parvoviridae and consist of nonenveloped icosahedral virions of 18–26 nm diameter, with a linear single-stranded DNA genome of 4680 nucleotides for the most characterized AAV2 strain [58]. The genome consists of two coding regions, *cap* and *rep*, which are flanked by ITRs at either end of the genome and an encapsidation signal (*Figure 3.4*). The *cap* gene encodes the capsid (coat) proteins and *rep* encodes for proteins involved in replication and integration functions [58]. AAV have been demonstrated to preferentially integrate into human chromosome 19 at site q13.4, directed by the *rep* genes, facilitating latent infection for the life of the cell [59]. AAV is, however, naturally replication incompetent and requires additional genes from a helper virus infection which in nature are generally complemented by adenovirus or herpes simplex virus co-infection [60].

AAV-based vectors generally involve replacement of the *rep* and *cap* genes with a transgene of interest, retaining the terminal repeats and packaging sequences essential to direct replication and packaging of the genome [58]. These vectors are therefore doubly helper-dependent requiring *trans*-complementation of the *rep* and *cap* functions, as well as the natural helper adenovirus functions [58]. Hence the applied *rep*-deleted AAV vectors cannot be rescued by adenovirus challenge alone, but also require wild-type AAV functions [61]. The nonpathogenic nature of AAV, having not been associated with any disease or tumor in humans, makes them potentially powerful gene therapy vehicles. *rep*-deleted AAV vectors have high transduction frequencies and have been reported to infect all cultured cells tested [58]. Systemic applications of recombinant AAV vectors *in vivo* have resulted in the transduction of the majority of tissues. Application of rAAV vectors to muscle has proved particularly successful

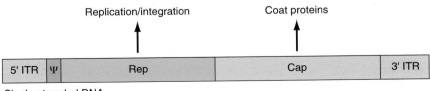

Figure 3.4. AAV genome. Schematic representation of the basic AAV genome, detailing the two major rep and cap coding units, flanked by the major regulatory inverted terminal repeat (ITR) elements and the packaging signal (Ψ).

resulting in significantly enhanced periods of expression and much reduced immunogenicity compared to other tissues [57].

The AAV vector system has a number of limitations at present. Although the preparation of titers of $>10^9$ iu/ml are feasible, the preparation is very laborious and problematic, with the stocks easily contaminated with wild-type AAV and helper viruses [58]. The toxicity associated with the *rep* proteins as well as the required helper adenovirus proteins, currently prevent the establishment of stable packaging cell lines. Additionally as the *rep* functions are deleted the assembled virions lose the ability to selectively integrate into human chromosome 19q13.4 and tend to integrate randomly, introducing the same risks of insertional mutagenesis as retroviruses [59]. One of the major limitations of AAV vectors is the limited insert capacity of approximately 4.7 kb [58].

3.5 Herpes simplex virus-based vectors

Herpes simplex virus (HSV) belongs to the herpesvirus family, a diverse group of large DNA viruses, all of which have the potential to establish lifelong latent infection [62]. HSV consists of 110 nm diameter particles comprising an icosahedral nucleocapsid, surrounded by a protein matrix, the tegument, which in turn is surrounded by a glycolipid-containing envelope [63]. The HSV genome consists of a linear double-stranded DNA molecule of 152 kb in length encoding 81 known genes, 38 of which are essential for virus production *in vitro* [63]. HSV has a wide tropism, infecting virtually any human cell, and is capable of nuclear delivery, infecting dividing as well as quiescent cells [63]. Herpesviruses infect and persist in cells of the nervous system, hence herpesvirus-based vectors may provide a unique strategy for gene transfer to cells of the nervous system [64].

Current HSV-based vectors involve replacement of one or more of the five immediate early (IE) genes whose functions are *trans*-complemented by packaging cell lines [63,65]. Titers of 10^8-10^9 viral particles per ml are possible with replication-competent HSV vectors, although the possible titers drop as more genetic sequence is deleted ($\sim 10^6$ particles/ml). Single IE gene deletion HSV vectors have been demonstrated to efficiently direct reporter gene expression in both myoblasts and differentiated myotubes in culture to similar degrees [63]. *In vivo* application to newborn mouse muscle was reported to result in significant numbers of positive fibers, but low level transduction was achieved in adult myofibers, believed to relate to HSV particles poorly penetrating the basal lamina [63]. The persistence of expression in both newborn and adult mice was, however, very limited which was believed to stem from cytotoxic effects resulting from expression of the remaining IE genes. Although preliminary results have indicated that the cytotoxic effects are reduced upon deletion of further IE genes, a great deal more research is required to refine HSV vectors including mechanisms of permeabilizing the basal lamina [63]. The major interest comes from the possibility of a gutless HSV vector, which has the potential of accommodating up to 150 kb of insert DNA [66].

3.6 Other viral vectors and hybrid vector systems

Despite the development of increasingly complex gene delivery systems, virus-mediated gene delivery remains the most efficient mechanism of gene transfer. Although Ad, RV, HSV and AAV are currently the most advanced gene therapy viral vector systems, a large number of other viruses do have potential. Specifically poliovirus, vaccinia virus, cyto-megalovirus (CMV), measles virus, semliki forest virus, sinbis virus, pseudorabies virus and parvovirus are being developed for gene therapy. No single viral vector or derivative system can, however, presently provide the necessary flexibility to the many applications of gene therapy proposed. The vast variability in host range and uptake potentials of the many tissues of the human body, as well as the many biological barriers to reaching the target tissues indicates that disease-specific gene targeting strategies will require the development of multiple gene delivery systems. Viral vectors are designed to utilize specific biological properties of viruses, such as cell receptor ligands for entry and mechanisms of integration into the host genome, that have evolved over time in relationship with the host. The natural response of the host, however, has also developed over millions of years to eliminate disease inducing viral infections. Current strategies of viral vector design are working to engineer viruses with predictable biological properties, maintaining the biological advantages of the virus that have been selected by nature while reducing the immunogeni-city of the viral components. Hybrid vectors are being developed which combine advantageous properties of current viral vector systems into single systems, generating novel viruses with unique combinations of functions aimed at attaining stable long-term expression in target cells.

Parvoviridae other than AAV are emerging as promising candidate gene transfer vehicles. Novel packaging strategies have recently been reported combining the salient features of the nonpathogenic parvovirus AAV-2 with the pathogenic parvovirus B19 which targets human hematopoietic cells in the erythroid lineage. Chimeric vectors were generated by packaging heterologous DNA sequences within the ITRs of AAV and subsequent packaging into the capsid structures of parvovirus B19 [67]. This vector was developed to achieve erythroid cell-specific delivery and expression of the transgene. Hence the development of parvovirus-based vectors would be particularly useful as potential treatments for hemoglobinopathies in general, and sickle cell anemia and β-thalassemia in particular [68]. Additionally, as $\sim 90\%$ of the human population are predicted to be seropositive for AAV capsid proteins, compared to $\sim 60\%$ for parvovirus B19 [67], hybrid vector systems could also offer mechanisms of evading humoral inactivation mechanisms.

Ad/AAV hybrid viruses have been constructed utilizing the infectivity, nuclear localization and ease of high titer preparation properties of Ad and the transgene amplification and integration properties of AAV. The Adenovirus/AAV hybrid consisted of a transgene cassette flanked by the AAV ITRs inserted into the E1 deleted region of an Ad5 genome [69]. Although the AAV *rep* gene could not be incorporated into the vector due to its inhibitory effect on adenovirus replication, it could be supplied as an extraviral component conjugated to the virions by a poly-L-lysine bridge. This adenovirus/AAV hybrid vector was demonstrated to efficiently deliver and rescue the AAV genome utilizing the infectivity and nuclear localization functions of the adenovirus. The AAV *rep* directed integration properties of this system, however, have yet to be assessed. A parallel AAV hybrid system has also been investigated utilizing the properties of HSV-I entry and nuclear localization, instead of Ad, together with the AAV properties of transgene amplification and integration into the host chromosome. A hybrid genome containing the HSV-I elements necessary to direct viral DNA replication and packaging into virions, as well as the additional AAV *rep* gene and ITRs to facilitate the AAV functions has been reported [70]. This vector has been demonstrated to efficiently infect quiescent as well as dividing cells both *in vitro* and *in vivo*, and was stably retained in a cell line in culture for more than 25 passages [71].

Baculoviruses (insect viruses) have recently been shown to have tropism for human hepatocytes, but are limited by transient transduction of primary cultures [72]. Bac/AAV hybrid viruses have since been constructed which associate the stable site-specific integration functions of AAV with the elevated transduction efficiencies and insert capacities of baculovirus [72]. These hybrid vectors have the potential to allow permanent, nontoxic gene delivery of DNA constructs for *ex vivo* treatment of primary human cells.

The HIV-I lentivirus has a limited host range and generates low titers, however, it can efficiently infect nondividing as well as dividing cells and stably integrate into the host genome [16]. Circumvention of these limiting factors has been investigated by substituting the HIV-I *env* and highly virulent accessory genes with the vesicular stomatitis virus (VSV) G glycoprotein, while retaining the HIV-I packaging sequences, and the genes for integration and nuclear localization. The resulting VSV pseudotyped lentiviruses were demonstrated to have a broad host range and stable physical properties which permit purification of high titer stocks [73]. Additional strategies of eliminating or substituting lentiviral accessory genes with other viral genes are also being investigated in the context of constructing safer, further attenuated, pseudotyped lentivirus vectors [74].

Chimeric vectors have been developed utilizing the favorable aspects of retroviral integration and adenoviral infectivity combined to achieve stable genetic transduction and efficient delivery *in vivo* [53,75]. Adenoviral vectors containing retroviral genome cassettes and the required packaging functions (*gag, pol* and *env*) enable the induction of retroviral producer cells *in vivo*, which can assemble infectious retroviral particles to infect neighbouring tissues [75]. An alternative strategy incorporating a retroviral cassette into an adenoviral vector which can be efficiently excised and subsequently integrated into the host genome following *in vivo* delivery of the adenoviral vector, together with the required excision/integration machinery, is also presenting great potential as a gene therapy vehicle [53]. Additional retroviral/poxviral chimeras have been reported, utilizing defective vaccinia as a chimeric carrier for retroviral genomes [76]. These chimeras enable efficient delivery of retroviral genomes to packaging cell lines significantly increasing the retroviral vector titers

Integration of viral genomes into host chromosomes with efficiencies possible with high titer viral stocks could pose significant risks of genotoxicity, due to insertional mutagenesis. The prospect of preferentially targeted integration at specific sites of the host genomes possible with AAV could also result in genotoxicity in certain cell types. Additionally, as the targeted integration is not 100%, with random integration possible in the absence of the AAV *rep* protein, extreme care needs to be taken in the application of these viruses, especially at high multiplicity of infection (MOI). The alternative of nonreplicating extrachromosomal genomes, characteristic of adenovirus and HSV vectors, are not stably maintained in dividing cells and are more suited to quiescent/nondividing cell populations, such as muscle or neuronal tissues. New generations of gene delivery vehicles are being developed which contain specific viral sequences which direct replication of the genome autonomously as episomal elements. In particular the Epstein–Barr virus (EBV) origin of replication (ori-P element) and the EBV nuclear antigen I (EBNA-I) gene, which have the potential to maintain vector DNA as autosomal extrachromosomal units in dividing cells, are being developed [77]. An HSV-I-based hybrid vector containing the ori-P element and EBNA gene has been demonstrated to maintain the hybrid vector extrachromosomally in culture for at least 6 months [77]. Hence this system presents great potential for gene transfer to dividing cells *in vivo*.

The construction of chimeric viruses is not a new technology and has been reported to occur in nature. Although human adenoviruses do not normally replicate in primate cells, upon coinfection with simian virus 40 (SV40) hybrid Ad/SV40 have been detected in nature [78]. The Ad genomes were determined to acquire sequences from the SV40 genomes (large T-antigen) which permitted their replication and assembly of hybrid genomes into wild-type Ad capsid particles. Hence the construction of hybrid viruses can be considered a natural course of evolution. While the hundreds of viruses determined to date have evolved over millions of years of evolution, scientists have now developed the technology to engineer viruses of their choice in the laboratory, massively accelerating evolution to the benefit of gene therapy. Care must, however, be taken to ensure that novel viruses are not generated which are many times more pathogenic than the wild-type viruses, such as a daunting Ebola/HIV/influenza hybrid virus!

References

1. Boris-Lawrie, K. and Temin, H.M. (1994) The retroviral vector–replication cycle and safety considerations for retrovirus-mediated gene-therapy. *Ann. New York Acad. Sci.*, **716**, 59–71.
2. Morgenstern, J.P. and Land, H. (1991) Choice and manipulation of retroviral vectors. In: *Methods in Molecular Biology, Vol. 7, Gene Transfer and Expression Protocols* (ed. E.J. Murray). The Human Press Inc., Clifton, NJ, pp. 181–205.
3. Coffin, J.M. (1996) Retroviridae: the viruses and their replication. In: *Fundamental Virology*, 3rd Edn (eds B.N. Field and D.M. Knipe), Raven Press, New York, pp. 771–813.
4. Wong-Staal, F. (1991) Human immunodeficiency viruses and their replication. *Fundamental Virology*, 2nd Edn (eds B.N. Fields and D.M. Knipe). Raven Press, New York, pp. 709–723.
5. Vile, R. (1991) The retroviral life cycle and the molecular construction of retroviral vectors. In: *Methods in Molecular Biology*, Vol. 8: *Practical Molecular Virology: Viral Vectors for Gene Expression.* (ed. M. Collins). The Humana Press, Inc., Clifton, NJ, pp. 1–15.
6. Miller, A.D. (1990) Retroviral packaging cells. *Hum. Gene Ther.*, **1**, 5–14.
7. Miller, A.D. Bender, M.A., Harris, E.A.S., Kaleko, M. and Gelinas, R.E. (1988) Design of retrovirus vectors for gene transfer and expression of the human β-globin gene. *J. Virol.*, **62**, 4337–4345.
8. Cosset, F.-L. and Russell, S.J. (1996) Targeting retrovirus entry. *Gene Therapy*, **3**, 946–956.
9. Miller, D.B., Adam, M.A. and Miller, A.D. (1990) Gene transfer by retrovirus vectors occurs only in *Cells* that are actively replicating at the time of infection. *Mol. Cell Biol.* **10**, 4239–4242.
10. Welsh, R.M., Cooper, N.R., Jensen, F.C. and Oldstone, M.B.A. (1975) Human serum lyses RNA tumour viruses. *Nature*, **257**, 612–614.
11. Cossett, F.-L., Takeuchi, Y., Battini, J.-L., Weiss, R.A. and Collins, M.K.L. (1995) High titre packaging cells producing recombinant retroviruses resistant to human complement. *J. Virol.*, **69**, 7430–7436.
12. Takeuchi, Y., Cosset, F.L.C., Lachmann, P.J., Okada, H., Weiss, R.A. and Collins, M.K.L (1994) Type-c retrovirus inactivation by human-complement is determined by both the viral genome and the producer cell. *J. Virol.*, **68**, 8001–8007.
13. Salvatori, G., Ferrari, G., Mezzogiorno, A., Servidei, S., Coletta, M., Tonali, P., Giavazzi, R., Cossu, G. and Mavilio, F. (1993) Retroviral vector-mediated gene transfer into human primary myogenic *Cells* leads to expression in muscle fibres *in vivo*. *Hum. Gene Ther.*, **4**, 713–722.
14. Fassati, A., Wells, D.J., Walsh, F.S. and Dickson, G. (1996) Transplantation of retroviral producer *Cells* for in-vivo gene-transfer into mouse skeletal-muscle. *Hum. Gene Ther.*, **7**, 595–602.
15. Jahner, D., Stuhlmann, H., Stewart, C.L., Harbers, K., Lohler, J., Simon, I. and Jaenisch, R. (1982) Denovo methylation and expression of retroviral genomes during mouse embryogenesis. *Nature*, **298**, 623–628.

16. Zufferey, R., Nagy, D., Mandel, R.J., Naldini, L. and Trono, D. (1997) Multiply attenuated lentiviral vector achieves efficient gene delivery *in vivo*. *Nat. Biotech.*, **15**, 871–875.

17. Johnston, J. and Power, C. (1999) Productive infection of human peripheral blood mononuclear cells by feline immunodeficiency virus: Implications for vector development. *J. Virol.*, **73**, 2491–2498.

18. Delviks, K.A., Hu, W.U. and Pathak, V.K. (1997) Psi(-) vectors: murine leukemia virus-based self-inactivating and self-activating retroviral vectors. *J. Virol.*, **71**, 6218–6224.

19. Olson, P., Temin, N.M. and Dornburg, A. (1992) Unusually high frequency of recombination of LTRs in U3-minus retrovirus vectors by DNA recombination or gene conversion. *J. Virol.*, **66**, 1336–1343.

20. Zhang, W.W. (1997) Review: Adenovirus vectors: development and application. *Exp. Opin. Invest. Drugs*, **6**, 1419–1457.

21. Horwitz, M.S. (1990) Adenoviridae and their replication. In: *Fundamental Virology*, 2nd Edn (eds B.N. Field, D.M. Knipe and R.M. Channock). Raven Press, New York, pp. 1679–1740.

22. Shenk, T. (1996) Adenoviridae: the viruses and their replication. In *Fields Virology*, Vol. 2, 3rd Edn (eds B.N. Fields, D.M. Knipe and P.M. Howley). Lippincott-Raven, Philadelphia, PA, pp. 2111–2148.

23. Norrby, E. (1969) The structural and functional diversity of Adenovirus capsid components. *J. Gen. Virol.*, **5**, 221–236.

24. Bergelson, J.M., Cunningham, J.A., Droguett, G., Kurtjones, E.A., Krithivas, A., Hong, J.S., Horwitz, M.S., Crowell, R.L. and Finberg, R.W. (1997) Isolation of a common receptor for coxsackie b viruses and adenoviruses 2 and 5. *Science*, **275**, 1320–1323.

25. Davidson, E., Diaz, R.M., Hart, I., Santis, G. and Marshall, J.F. (1997) Integrin α5β1-mediated adenovirus infection is enhanced by the integrin-activating antibody TS2/16. *J. Virol.*, **71**, 6204–6207.

26. Grable, M. and Hearing, P. (1992) *Cis* and *trans* requirements for the selective packaging of adenovirus type-5 DNA. *J. Virol.*, **66**, 723–731.

27. Bett, A.J., Prevec, L. and Graham, F.L. (1993) Packaging capacity and stability of human adenovirus type 5 vectors. *J. Virol.*, **67**, 5911–5921.

28. Graham, F.L., Smiley, J., Russell, W.C. and Nairn, R. (1977) Characterisation of a human cell line transformed by DNA from human adenovirus type 5. *J. Gen. Virol.*, **36**, 59–74.

29. Bett, A.J., Haddara, W., Prevec, L. and Graham, F.L. (1994) An efficient and flexible system for construction of adenovirus vectors with inserts or deletions in early regions 1 and 3. *Proc. Natl Acad. Sci. USA*, **91**, 8802–8806.

30. Berkner, K.L. (1988) Development of adenovirus vectors for the expression of heterologous genes. *BioTechniques*, **6**, 616–629.

31. Kass-Eisler, A., Falck-Pedersen, E., Elfenbein, D.H., Alvira, M., Buttrick, P.M. and Leiwand, L.A. (1994) The impact of developmental stage, route of administration and the immune system on adenovirus-mediated gene transfer. *Gene Ther.*, **1**, 395–402.

32. Rosenfeld, M.A., Siegfried, W., Yoshimura, K. *et al.* (1991). Adenovirus-mediated transfer of a recombinant α1-antitrypsin gene to the lung epithelium *in vivo*. *Science*, **252**, 431–434.

33. Horwitz (1990) Adenoviruses. In: *Fundamental Virology*, 2nd Edn (ed. B.N. Field, D.M. Knipe and R.M. Channock). Raven Press, New York, pp. 1723–1740.

34. Yang, Y., Nunes, F.A., Berencsi, K., Gonczol, E., Engelhardt, J.F. and Wilson, J.M. (1994) Inactivation of E2A in recombinant adenoviruses improves the prospect for gene-therapy in cystic-fibrosis. *Nat. Genet.*, **7**, 362–369.

35. Grubb, B.R., Pickles, R.J., Ye, H., Yankaskas, J.R., Vick, Rn., Engelhardt, J.F., Wilson, J.M., Johnson, L.G. and Boucher, R.C. (1994) Inefficient gene-transfer by adenovirus vector to cystic-fibrosis airway epithelia of mice and humans. *Nature*, **371**, 802–806.

36. Yang, Y., Jooss, K.U., Su, Q., Ertl, H.C.J. and Wilson, J.M. (1996) Immune response to viral antigens versus transgene product in the elimination of recombinant adenovirus-infected hepatocytes *in vivo*. *Gene Ther.*, **3**, 137–144.

37. Ragot, T., Vincent, N., Chafey, P., Vigne, E., Gilgenkrantz, H., Couton, D., Cartaud, J., Briand, P., Kaplan, J.C., Perricaudet, M. and Kahn, A. (1993) Efficient adenovirus-mediated transfer of a human mini-dystrophin gene to skeletal muscle of mdx mice. *Nature*, **361**, 647–650.

38. Yei, S., Mittereder, N., Te, K., O'Sullivan, C. and Trapnell, B.C. (1994) Adenovirus-mediated gene transfer for cystic fibrosis: quantitative evaluation of repeated *in vivo* vector administration to the lung. *Gene Ther*, **1**, 192–200.

39. Dong, J.-Y., Wang, D., Van Ginkel, F.W., Pascual, D.W. and Frizzell, R.A. (1996) Systematic analysis of repeated gene delivery into animal lungs with a recombinant adenovirus vector. *Hum. Gene Ther.*, **7**, 319–331.

40. Zhou, H.S., O'Neal, W., Morral, N. and Beaudet, A.L. (1996) Development of a complementing cell-line and a system for construction of adenovirus vectors with E1 and E2A deleted. *J. Virol.*, **70**, 7030–7038.

41. Gorziglia, M.I., Kadan, M.J., Yei, S., Lim, J., Lee, G.M., Luthra, R. and Trapnell, B.C. (1996) Elimination of both E1 and E2a from adenovirus vectors further improves prospects for *in vivo* human gene therapy. *J. Virol.*, **70**, 4173–4178.

42. Wang, Q., Greenburg, G., Bunch, D., Farson, D. and Finer, M.H. (1997) Persistent transgene expression in mouse liver following *in vivo* gene transfer with a ΔE1/ΔE4 adenovirus vector. *Gene Ther.*, **4**, 393–400.

43. Armentano, D., Zabner, J., Sacks, C., Sookdeo, C.C., Smith, M.P., St. George, J.A., Wadsworth, S.C., Smith, A.E. and Gregory, R.J. (1997) Effect of the E4 region on the persistence of transgene expression from adenovirus vectors. *J. Virol.*, **71**, 2408–2416.

44. Beier, D.C., Cox, J.H., Vining, D.R., Cresswell, P. and Engelhard, V.H. (1994) Association of human class I MHC alleles with the adenovirus E3/19K protein. *J. Immunol.*, **152**, 3862–3872.

45. Dai, Y., Schwarz, E., Gu, D., Zhang, W., Sarvetnick, N. and Verma, I. (1995). Cellular and humoral immune responses to adenoviral vectors containing factor IX gene: Tolerization of factor IX and vector antigens allows for long-term expression. *Proc. Natl Acad. Sci. USA*, **92**, 1401–1405.

46. Engelhardt, J.F., Ye, X., Doranz, B. and Wilson, J.M. (1994) Ablation of E2A in recombinant adenoviruses improves transgene persistence and decreases inflammatory response in mouse liver. *Proc. Natl Acad. Sci. USA*, **91**, 6196–6200.

47. Lochmuller, H., Petrof, B.J., Pari, G. *et al.* (1996) Transient immunosuppression by FK506 permits a sustained high-level dystrophin expression after adenovirus-mediated dystrophin minigene transfer to skeletal-muscles of adult dystrophic (mdx) mice. *Gene Ther.*, **3**, 706–716.

48. Kaplan, J.M. and Smith, A.E. (1997) Transient immunosuppression with deoxyspergualin improves longevity of transgene expression and ability to readminister adenoviral vector to the mouse lung. *Hum. Gene Ther.*, **8**, 1095–1104.

49. Kumar-Singh, R. and Chamberlain, J.S. (1996) Encapsidated adenovirus minichromosomes allow delivery and expression of a 14kb dystrophin cDNA to muscle cells. *Hum. Mol. Genet.*, **5**, 913–921.
50. Chen, H.H., Mack, L.M., Kelly, R., Ontell, M., Kochanek, S. and Clemens, P.R. (1997) Persistence in muscle of an adenoviral vector that lacks all viral genes. *Proc. Natl Acad. Sci. USA*, **94**, 1645–1650.
51. Parks, R.J., Chen, L., Anton, M., Sankar, U., Rudnicki, A. and Graham, F.L. (1996) A helper-dependent adenovirus vector system: Removal of helper virus by Cre-mediated excision of the viral packaging signal. *Proc. Natl Acad. Sci. USA*, **93**, 13565–13570.
52. Haecker, S.E., Stedman, H., Balice-Gordan, R.J., Smith, D.B.J., Greelish, J.P., Mitchell, M.A., Wells, A., Sweeney, H.L. and Wilson, J.M. (1996) *In vivo* expression of full-length dystrophin from adenoviral vectors deleted of all viral genes. *Hum. Gene Ther.*, **7**, 1907–1914.
53. Murphy, S.J., Bell, S.J., Chong, H., Diaz, R.M. and Vile, R.G. (1998) Adenovirus-mediated delivery of an excisable retroviral cassette which can stable integrate into target cell genomes. *J. Gene Med.*, **1**, Suppl. 74, Abstract.
54. Parks, R.J. and Graham, F.L. (1997) A helper-dependent system for adenovirus vector production helps define a lower limit for efficient DNA packaging. *J. Virol.*, **71**, 3293–3298.
55. Wickham, T.J., Carrion, M.E. and Kovesdi, I. (1995) Targeting of adenovirus penton to new receptors through replacement of its RGD motif with other receptor specific peptide motifs. *Gene Ther.*, **2**, 750–756.
56. Kransnykh, V., Dmitriev, I., Mikheeva, G., Miller, C.R., Belousova, N. and Curiel, D.T. (1998) Characterisation of an adenovirus vector containing a heterologous peptide epitope in the H1 loop of the fibre knob. *J. Virol.*, **72**, 1844–1852.
57. Monahan, P.E., Samulski, R.J., Tazelaar, J., Xiao, X., Nicholas, T.C., Bellinger, D.A., Read, M.S. and Walsh, C.E. (1998) Direct intramuscular injection with recombinant AAV vectors results in sustained expression in a dog model of haemophilia. *Gene Ther.*, **5**, 40–49.
58. Kremer, E.J. and Perricaudet, M. (1995) Adenovirus and adenoassociated virus-mediated gene-transfer. *British Med. Bull.*, **51**, 31–44.
59. Muzyczka, N. (1992) Use of adeno-associated virus as a general transduction vector for mammalian cells. *Curr. Top. Microbiol. Immunol.*, **158**, 97–129.
60. Verma, I.M. and Somia, N. (1997) Gene therapy-promises, problems and prospects. *Nature*, **389**, 239–242.
61. Bordignon, C., Notarangelo, L.D., Nobili, N. *et al.* (1995) Gene therapy in peripheral-blood lymphocytes and bone-marrow for ADA(-) immunodeficient patients. *Science*, **270**, 470–475.
62. Efstathiou, S. and Minson, A.C. (1995) Herpes virus-based vectors. *Brit. Med. Bull.*, **51**, 45–55.
63. Huard, J., Krisky, D., Oligino, T., Marconi, P., Day, C.S., Watkins, S.C. and Glorioso, J.C. (1997) Gene transfer to muscle using herpes simplex virus-based vectors. *Neuromusc. Disord.*, **7**, 299–313.
64. Hermens, W.T. and Verhaagen, J. (1998) Viral vectors-tools for gene transfer in the nervous system. *Progress Neurobiol.*, **55**, 399–432.
65. Brehm, M, Samaniego, L.A., Bonneau, R.H., DeLuca, N.A. and Tevethia, S.S. (1999) Immunogenicity if herpes simplex virus type 1 mutants containing deletions in one or more of the α-genes: ICP4, ICP27, ICP22, and ICP0. *Virology*, **256**, 258–269.

66. Frenkel, N., Singer, O. and Kwong, A.D. (1994) Minireview: The herpes simplex virus amplicon—a versatile defective virus vecto. *Gene Ther.*, **1**, S40–S46.
67. Ponnazhagan, S., Weigel, K.A., Raikwar, S.P., Mukherjee, P., Yoder, M.C. and Srivastava, A. (1998) Recombinant human parvovirus B19 vectors: erythroid cell-specific delivery and expression of transduced genes. *J. Virol.*, **72**, 5224–5230.
68. Srivastava, A., Wang, X.S., Ponnazhagan, S., Zhou, S.Z. and Yoder, M.C. (1996) Adeno-associated virus 2-mediated transduction and erythroid lineage-specific expression in human hematopoietic progenitor cells. *Curr. Top. Microb. Immun.*, **218**, 93–117.
69. Fisher, K.J., Choi, H., Burda, J., Chen, S.J. and Wilson, J.M. (1996) Recombinant adenovirus deleted of all viral genes for gene-therapy of cystic-fibrosis. *Virology*, **217**, 11–22.
70. Johnston, K.M., Jacoby, D., Pechan, P.A., Fraefel, C., Borghesani, P., Schuback, D., Dunn, R.J., Smith, F.I. and Breakefield, X.O. (1997) HSV/AAV hybrid amplicon vectors extend transgene expression in human glioma cells. *Hum. Gene Ther.* **8**, 359–370.
71. Fraefel, C., Jacoby, D.R., Lage, C., Hilderbrand, H., Chou, J.Y., Alt, F.W., Breakefield, X.O. and Majzoub, J.A. (1997) Gene transfer into hepatocytes mediated by helper virus-free HSV/AAV hybrid. *Mol. Med.*, **3**, 813–825.
72. Palombo, F., Monciotti, A., Recchia, A., Cortese, R., Ciliberto, G. and La Monica, N. (1998) Site-specific integration in mammalian cells mediated by a new hybrid baculovirus-adeno-associated virus vector. *J. Virol.* **72**, 5025–5034.
73. Naldini, L., Blomer, U., Gage, F.H., Trono, D. and Verma, I.M. (1996) Efficient transfer, integration, and sustained long-term expression of the transgene in adult rat brains injected with a lentiviral vector. *Proc. Natl Acad. Sci. USA.*, **93**, 11382–11388.
74. Gasmi, M.,. Glynn, J.,. Jin, M.J., Jolly, D.J., Yee, J.K. and Chen, S.T. (1999) Requirements for efficient production and transduction of human immuno-deficiency virus type 1-based vectors. *J. Virol.* **73**, 1828–1834.
75. Feng, M., Jackson, W.H., Goldman, C.K., Rancourt, C., Wang, M., Dusing, S.K., Siegal, G. and Curiel, D.T. (1997) Stable *in vivo* gene transduction via a novel adenoviral/retroviral chimeric vector [see comments]. *Nat. Biotech.*, **15**, 866–870.
76. Holzer, G.W., Mayrhofer, J.A., Gritschenberger, W., Dorner, F. and Falkner, F.G. (1999) Poxviral/retroviral chimeric vectors allow cytoplasmic production of transducing defective retroviral particles. *Virology*, **253**, 107–114.
77. Wang, S. and Vos J.M. (1996) A hybrid herpesvirus infection vector based on Epstein–Barr virus and herpes simplex virus type I for gene transfer into human cells *in vitro* and *in vivo*. *J. Virol.*, **70**, 8422–8430.
78. Lewis, A.M. (1998) SV40 in adenovirus vaccines and adenovirus-SV40 recombinants. *Dev. Biol. Stand*, **94**, 207–216.

Chapter 4

Nonviral delivery systems for gene therapy

Andrew D. Miller

4.1 Introduction

Nonviral delivery systems (or vectors) are in some ways the poor relation to viral delivery systems dealt with in the previous chapter, but that is not to say that they may not rival or even surpass viral delivery systems in time. Given the fact that millions of years of evolution have been devoted to perfecting the art of viral gene delivery, it should not come as much of a surprise that the efficiency of nonviral delivery systems is not currently that competitive with the efficiency of viral systems. However, this chapter will aim to illustrate that in spite of this current state of affairs, nonviral delivery systems are still proving effective agents of nucleic acid delivery and given further development may well be the pharmaceutical delivery agents of choice for gene therapy in the future. There has been a general tendency to consider nonviral delivery systems to include all chemical and physical methods for nucleic acid delivery aside from virus-based methods. However, for the purposes of this chapter, I will confine most of my remarks to the artificial chemical systems which compare more directly with viral delivery systems and are more likely to form clinically useful delivery agents in the future.

4.1.1 Basic principles of chemical nonviral delivery systems

The vast majority of current chemical nonviral delivery systems are based around polycationic entities which cause compaction of negatively charged nucleic acids accompanied by the formation of nanometric complexes. Typically, these polycationic entities belong to one of two main categories which are either *cationic liposome/micelle* or *cationic polymer*-based. *Lipoplex* is the name given to the nanometric complex

43

formed between a cationic liposome/ micelle and nucleic acids; *polyplex* the name given to the nanometric complex formed between a cationic polymer and nucleic acids [1]. Once formed, these nanometric complexes are generally stable enough to protect bound nucleic acids from degradation and are competent to enter cells, usually by endocytosis (*Figure 4.1*). The general presumption is that these complexes need to have a net positive charge to enter cells and that endocytosis is triggered by nonspecific interaction between cationic complexes and the anionic cell-surface proteoglycans of adherent cells. Once inside, a proportion of the bound nucleic acids should be able to dissociate and escape from early-endosomes into the cytoplasm either to perform a therapeutic function there as in the case of mRNA (Path B; *Figure 4.1*), or else traffic into the nucleus to perform a therapeutic function, as in the case of DNA (Path C; *Figure 4.1*).

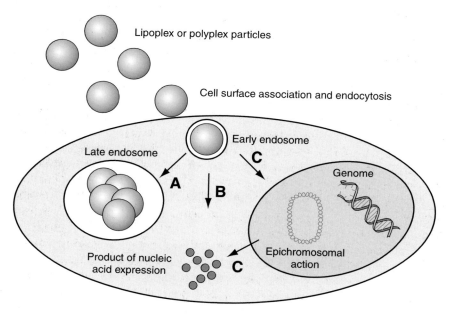

Figure 4.1. Schematic diagram to show the process by which chemical nonviral delivery systems deliver nucleic acids to cells. Lipoplex (or polyplex) particles which have not succumbed to aggregation and/or serum-inactivation (as appropriate), associate with the cell surface and enter usually by endocytosis. The majority of lipoplex (or polyplex) particles in early endosomes become trapped in late endosomes (Path A) and the nucleic acids fail to reach the cytosol. A minority of lipoplex (or polyplex) particles in early endosomes are able to release their bound nucleic acids into the cytosol. Path B is followed by RNA which may act directly in the cytosol. Path C is followed by DNA which must enter the nucleus in order to act. The diagram is drawn making the assumption that plasmid DNA has been delivered which is expressed in an epichromosomal manner.

4.1.2 General problems for chemical nonviral delivery systems

Every stage of the mechanism outlined in the previous section (*Figure 4.1*) is attended with many problems. All these problems have been well documented for cationic liposome/micelle-based systems [2,3] and also for cationic polymer-based nonviral delivery systems (see *Section 4.3*). In the first instance, lipoplexes or polyplexes frequently show instability towards aggregation even before entering a cell, especially under conditions of physiological salt levels. Furthermore, they are also unstable in the presence of body fluid components such as serum proteins. In addition, current lipoplex and to some extent polyplex mixtures are frequently not homogeneous and monodisperse (as depicted in *Figure 4.1*) but are instead polydisperse, a characteristic which is hardly likely to be beneficial either. Without additional help, neither lipoplexes nor polyplexes are especially cell-type selective. Similarly, without additional help, neither will enter cells very rapidly (uptake time is hours). Even once inside a cell, problems are far from over. The vast majority of lipoplex and probably polyplex particles will aggregate within late-endosome compartments (Path A; *Figure 4.1*) producing macroparticles from which nucleic acids only dissociate and escape with difficulty. Moreover, if nucleic acids do manage to escape the endosome compartments after delivery, access to the nucleus may still represent yet another considerable barrier to nucleic acid function.

Hence in truth, nucleic acid delivery by chemical nonviral delivery systems is very inefficient. However, as the following sections will make clear, in spite of this inefficiency there have still been some impressive results obtained using nonviral delivery systems to introduce genes into cells *in vivo*, and even in a clinical context. Unfortunately, as the following sections will also make clear, much of the previous and current research into chemical nonviral delivery systems has been mostly concerned with the identification of new and different polycationic entities capable of forming complexes with nucleic acids and delivering the nucleic acids to cells *in vitro*. Accordingly, there has been only a little systematic research carried out to date to solve the above problems for any one chemical nonviral delivery system. As a result, most of the current systems discussed are unlikely to have any ultimate clinical use. Nevertheless, systematic research programs are now beginning and solutions to many of the problems described will be found in time.

4.2 Cationic liposome/micelle-based nonviral delivery systems

Cationic liposome/micelle-based systems have come closer to providing clinically effective gene delivery than any other chemical nonviral delivery systems to date. These systems are formed from either a single synthetic cationic amphiphile (known as a cytofectin; *cyto-* for cell and-*fectin* for transfection [i.e. gene delivery and expression]) or more commonly from the combination of a cytofectin and a neutral lipid such as dioleoyl L-α-phosphatidylethanolamine (DOPE). Currently, there are at least 30 new cationic liposome/micelle systems which have been reported to mediate nucleic acid delivery to cells, of which a number have been commercialized (*Table 4.1*) [4–34]. More are being reported all the time, but in each case the key ingredient is the cytofectin used. The structures of a number of representative cytofectins are shown (*Figure 4.2*). Typically, cytofectin and neutral lipid components are mixed together in an appropriate mol ratio and then induced or formulated into unilammellar vesicles by any one of a number of methods including reverse phase evaporation (REV), dehydration–rehydration (DRV), extrusion, microfluidization, etc (see *Table 4.1*). Alternatively, cytofectins may be assembled into micellar structures after being dispersed in water or aqueous organic solvents (see *Table 4.1*). Unilammellar vesicles or micelles may then be combined with nucleic acids to form lipoplexes which are able to deliver nucleic acids into cells. Cationic liposome/micelle-mediated delivery of nucleic acids is optimal *in vivo* when the the mol ratio of cationic liposome/micelle-to-nucleic acid in the lipoplex mixture is such that the positive/negative charge ratio is around 1 or greater [35–37]. *In vitro* the optimal positive/negative charge ratio may be much higher than 1 [8,38], but more generally the optimal ratio is closer to 1 [14,39–41]. Under circumstances where comparisons have been made between lipoplex structures and efficiency of nucleic acid delivery *in vitro*, evidence suggests that optimal lipoplex mixtures are in fact heterogeneous and polydisperse, being comprised of a variety of structures all in dynamic equilibrium [42,43]. These structures have been variously identified and described by a number of researchers and they include multilammellar lipid/nucleic acid clusters (>100 nm in diameter) [44–46] perhaps with some surface associated nucleic acids [47], thinly lipid-coated nucleic acid strands [48] and free nucleic acids [45]. Which of this polydisperse mixture is responsible for actual nucleic acid delivery is not entirely clear!

One of the greatest problems that researchers have experienced in studying cationic liposome/micelle mediated nucleic acid delivery has been the consistent observation that the *in vitro* efficacy of a given cationic liposome/micelle system is a poor guide to the *in vivo* efficacy [24,25]. Consequently, considerable resources have had be devoted to *in vivo* tests, frequently with little meaningful guidance from *in vitro* data. This represents a serious problem especially given the fact that the vast majority of research into cationic liposome/micelle systems has been concerned with *in vitro* delivery of nucleic acids [2]. The most directed attempts to use cationic liposome/micelle systems for gene therapy research has been in the sphere of nucleic acid delivery to the lung. In particular, research has been motivated by an interest in developing non-viral gene therapy approaches for the treatment of cystic fibrosis.

Table 4.1. Main cationic liposome/micelle systems to date.

Cytofectin	Formulation	Trade name/manufacturer	Reference
Commercialized			
DOTMA	DOTMA/DOPE 1:1 (w/w)	Lipofectin/GIBCO BRL	[4]
DOTAP	DOTAP	DOTAP/Roche Molecular	[5]
DOSPA	DOSPA/DOPE 3:1 (w/w)	LipofectAMINE/GIBCO BRL	[6]
Tfx	Tfx/DOPE (Tfx-10, -20, -50)	Tfx™/Promega	[7]
DOGS	micelle	Transfectam/Promega	[8]
Di C 14 amidine	Di C 14 amidine/DOPE 1:1 (m/m)	Clonfectin/Clontech	[9,10]
DDAB	DDAB/DOPE 1:2.5 (w/w)	LipofectACE/GIBCO BRL	[11]
Cholesteryl-spermidine	Cholesteryl-spermidine	Transfectall/Apollon Inc.	[12]
DC-Chol	DC-Chol/DOPE 6:4 (m/m)	DC-Chol/Sigma	[13]
DOSPER	DOSPER	DOSPER/Roche Molecular	[33]
Not commercialized			
DMRIE	DMRIE/DOPE 1:1 (m/m)		[14]
DORIE	DORIE/DOPE 1:1 (m/m)		[14]
GAP-DLRIE	GAP-DLRIE/DOPE 1:1 (m/m)		[15]
DORI	DORI/DOPE 1:1 (m/m)		[14,16]
14 Dea 2	14 Dea 2		[17,18]
GS 2888	GS 2888/DOPE 1:1 (m/m)		[19]
RPR 120535	micelle		[20,21]
DLS	DOGS/DOPE 6:10 (m/m)		[22]
Cholic acid hexamine	Cholic acid hexamine/DOPE 1:1 (w/w)		[23]
GL-67	GL-67/DOPE 1:2 (m/m)		[24]
CTAP	CTAP/DOPE 1:2 (m/m)		[25]
BGTC	BGTC/DOPE 3:2 (m/m)		[26]
Lys-Pam$_2$-GroPEtn	Lys-Pam$_2$-GroPEtn/Chol/eggPC 1.5:3.0:5.5(w/w/w)		[27]
L-PE	L-PE/CEβA 6:4 (m/m)		[28]
EDMPC	EDMPC/DOPE 1:1 (m/m)		[29]
$2C_{12}$-L-Glu-ph-C_2-N$^+$	$2C_{12}$-L-Glu-ph-C_2-N$^+$		[30]
DOTIM	DOTIM/Chol 1:1 (m/m) or DOTIM/DOPE 1:1 (m/m)		[31]
SAINT	SAINT/DOPE 1:1 (w/w)		[32]
DODAC	DODAC/DOPE 1:1 (m/m)		[34]

Abbreviations; w/w: weight ratio; m/m: mol ratio. For chemical abbreviations see p. xi.

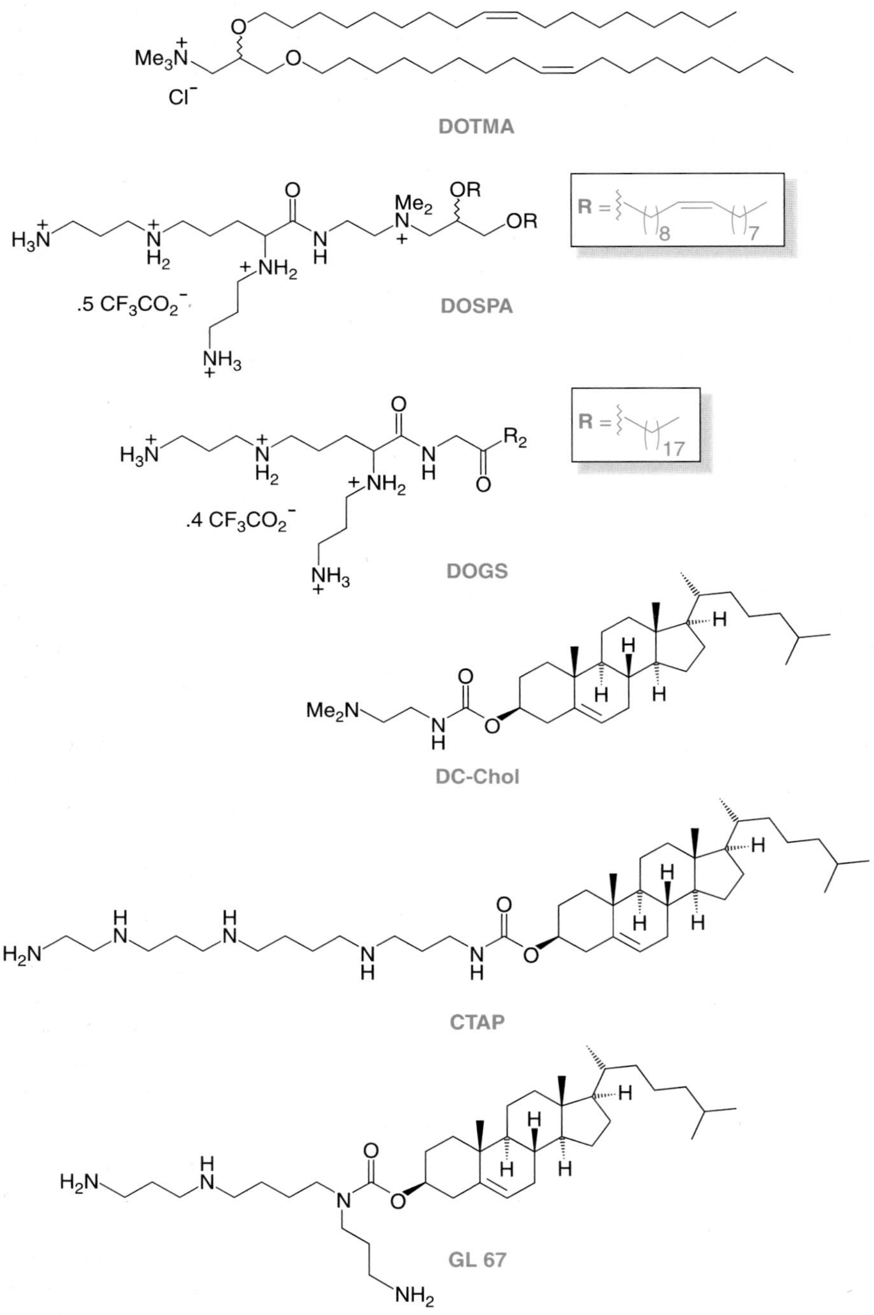

Figure 4.2. Some representative structures of cytofectins involved in cationic liposome/micelle systems. Details about the use of these various cytofectins may be found in *Table 4.1* and in the text.

Following a demonstration in 1992 by Debs and coworkers that DOPE/ *N*-[1-(2,3-dioleyloxy)propyl]-*N*,*N*,*N*-trimethyl ammonium chloride (DOT-MA) cationic liposomes could deliver a marker gene to murine lung by aerosol administration of the complex [49], Alton *et al.* were able to demonstrate that DOPE/3β-[*N*-(*N'*,*N'*-dimethylaminoethane)carbamoyl]-cholesterol (DC-Chol) liposomes could be used in a similar way to deliver the cystic fibrosis transmembrane conductance regulator (CFTR) gene to the lung epithelial cells of cystic fibrosis transgenic mice resulting in correction of lung defects associated with cystic fibrosis [39]. Subsequently, DOPE/DC-Chol liposomes were used in preparatory clinical trials directed at the human gene therapy of cystic fibrosis [50]. However, these experiments indicated that this first generation liposome system was unlikely to be efficient enough for general use in human lung gene therapy. Therefore, a number of attempts were made to derive improved, second generation cationic liposomes. The most successful to emerge from this process have been DOPE/GL-67 and DOPE/N^{15}-cholesteryloxycarbonyl-3,7,12-triazapentadecane-1,15-diamine (CTAP) liposome systems [24,25], both of which were found to be at least 100-fold more efficient at mediating gene delivery *in vivo* than DOPE/DC-Chol liposomes, therefore equivalent in efficacy to a low titer of adenovirus. A sterically stabilized formulation of the DOPE/GL-67 liposome system [GL-67/DOPE/ dimyristoyl L-α-phosphatidylethanolamino-poly(ethylene glycol)-5000 (DMPE-PEG$_{5000}$)] has since been used recently in a cystic fibrosis clinical trial and found to mediate a 25% correction of the basic chloride ion defect in the lungs of cystic fibrosis patients in comparison to the placebo [51]. Steric stabilization was necessary here to prevent aggregation of lipoplex mixtures at the concentrations required for successful *in vivo* delivery. Further improvements in gene delivery efficiency will probably be necessary for cystic fibrosis gene therapy to become clinical reality, but these results are an impressive step forward.

However, the preceding results could give a misleading impression if taken in isolation. Much less impressive results have been obtained with systemic and neurological applications of cationic liposome/micelle-based nucleic acid delivery. Typically, lipoplex mixtures, whether introduced by intraperitoneal (i.p.) or intravenous (i.v.) methods, will give widespread systemic delivery of nucleic acids most especially to the lung, spleen and kidney [52–55]. This demonstrable lack of cell and organ type specificity may be less of a problem in neurological tissue [56], but is still a real cause for concern where cell-specific delivery will be necessary for most clinical gene therapy applications. As a result, local administration and direct injection approaches have been used to try and circumvent this problem [57–59]. However, given the susceptibility of lipoplexes to aggregation at the higher concentrations required for *in vivo* nucleic acid delivery, even this approach is not without its problems [60]. In consequence, there has started to be some concerted attempts to overcome one or more of the problems outlined earlier (*Section 4.1.2*) in the hope that cationic liposome/micelle systems may be devized which will be far more efficacious for systemic and neurological delivery of nucleic acids *in vivo*. For instance, cationic liposomes sterically stabilized with poly(ethylene glycol)–phospholipid conjugates have been introduced to overcome the problem of lipoplex aggregation at higher concentrations and under systemic conditions of physiological salt levels [61]. Other methods of steric stabilization have also been described using detergents [36,62]. Such steric stabilization has the added bonus of reducing serum instability as well [36], presumably by sterically blocking the interaction of negatively charged serum proteins with the lipoplex surface, an interaction which is known to be the basis of serum-inactivation of lipoplexes [37]. Furthermore, adverse immune and other inflammatory effects, which may be a serious problem for systemic cationic liposome/micelle-mediated nucleic acid delivery *in vivo* [63,64], may also be usefully attenuated by steric stabilizing agents. There has been some suggestion that serum-inactivation effects may be simply overcome by increasing the cationic liposome/micelle-to-nucleic acid mol ratio in the lipoplex mixture such that the positive/negative charge ratio is significantly greater than 1 [37], or by introducing a lipoplex time-dependent maturation [65]. Both approaches could be of general use to improve *in vivo* applications provided they are both used in combination with other approaches. Finally, lipoplex stability *in vivo* may well also be enhanced by careful choice of neutral lipid in the formulation of cationic liposome/micelle systems. Whilst DOPE has been used traditionally, there is an increasing suggestion that cholesterol (Chol) may be a more appropriate neutral lipid for incorporation into lipoplexes applied systemically *in vivo* [61,66,67].

In addition to these innovations aimed at solving the basic problems of lipoplex susceptibility to aggregation and serum effects, there has been some progress with innovations to improve the cell-type selectivity of lipoplexes and their rate of cell entry. For instance, *in vitro* studies have been performed using lipoplexes containing triantennary galactolipids to target hepatoma cells [38], or alternatively covalently attached E-selectin selective murine monoclonal antibodies to target activated vascular human umbilical vein endothelial cells (HUVEC) [68]. These point the way to *in vivo* applications of such ligand-modified lipoplex systems, though in these and other reported cases there appears to be strong competition between the ligand-mediated process of cell binding and endocytosis and the nonspecific, more general interaction between lipoplex and anionic cell membranes [38,68,69]. This will need to be resolved properly if ligand-mediated effects are to have widespread application *in vivo*. Innovations to improve the escape of nucleic acids from endosomes following lipoplex cell entry have involved the use of low pH-activated membrane-active peptides such as GALA, a 30 amino acid, pH-sensitive α-helical amphiphilic peptide which induces membrane fusion and permeabilization at endosome acidic pH values [69]. The use of other amphiphilic peptides has also been described for this purpose [70]. Alternatively, pH-sensitive cationic liposomes have now been developed to both bind nucleic acids efficiently but also buffer endosomes [71]. In buffering endosomes, they presumably prevent nucleic acid degradation by inhibiting endosomal acidification [72], and subsequently induce osmotic swelling and endosome membrane rupture allowing nucleic acids to escape into the cytosol [73].

Recently, there has been a growing trend to move away from lipoplex to hybrid lipoplex systems which are showing some promise. One of the most interesting has been the formulation of lipid-protamine-DNA (LPD) complexes [74,75]. In these complexes, the peptide protamine supplements the lipoplex interaction between cationic liposome and DNA to produce dense spherical particles (mean diameter 135 nm) which are apparently much less heterogeneous and polydisperse than normal lipoplex mixtures. Although LPD complexes might well be expected to be susceptible to aggregation and serum effects given their high charge, they still give impressive levels of *in vivo* expression following systemic administration [74,75]. Such LPD or LPD-like systems could prove very potent in future applications. The recently described hemagglutinating virus of Japan (HVJ; Sendai virus)-cationic liposome system is a fascinating variation on the the LPD concept (*Figure 4.3*) [76]. HVJ-liposomes have been known for a little while and are something of a 'viral, nonviral hybrid' vector system, being a fusion of UV-irradiated virions of the HVJ and liposomes in which are encapsulated nucleic acids complexed with the High Mobility Group 1 (HMG-1) protein [76]. The HMG-1 protein is there to assist nuclear access and localization of delivered nucleic acids as well as promoting gene stabilization within the nuclear envelope [77]. Although negatively charged and approximately 350–500nm in size, HVJ-liposomes themselves have been used in an impressive range of *in vivo* systemic and anti-cancer applications [76]. One major reason for this success is the presence of the H_N and F proteins in the liposome bilayer (*Figure 4.3*). These proteins allow HVJ-liposomes to interact with cell surface sialic residues, fuse with the cell membrane and then release encapsulated nucleic acids directly into the cytoplasm, bypassing endocytosis altogether [76]. This neatly overcomes the problem of endosome escape that so confounds other chemical nonviral delivery systems. The corresponding development of the HVJ-cationic liposome system is only a recent phenomenon, and the *in vivo* potential of this system has yet to be fully investigated. Nevertheless, nucleic acid delivery by DC-Chol based HVJ-cationic liposomes to various mammalian cell types has proved very promising *in vitro* although rather less so *in vivo* [78]. In contrast, HVJ-cationic liposomes prepared with the cytofectin *N*-(α-trimethylammo-nioacetyl)-didodecyl-D-glutamate chloride (TMAG) have proved able to mediate delivery of nucleic acids to tracheal and bronchiolar epithelial

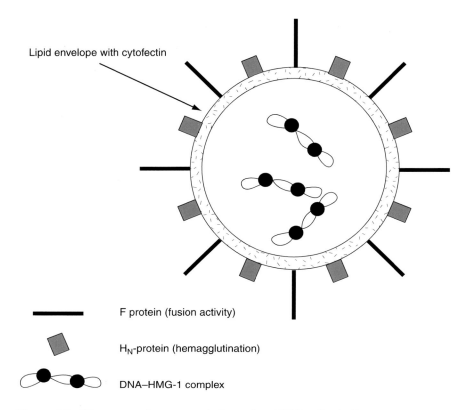

Lipid envelope with cytofectin

—————— F protein (fusion activity)

H$_N$-protein (hemagglutination)

DNA–HMG-1 complex

Figure 4.3. Diagrammatic representation of an HVJ-cationic liposome.

cells *in vivo* with reasonable efficiency [79]. Hence, given several inbuilt advantages alluded to earlier, HVJ-cationic liposomes could yet turn out to be very potent agents for the delivery of nucleic acids in gene therapy applications of the future.

4.3 Cationic polymer-based nonviral delivery systems

The application of cationic polymers to gene therapy research has lagged a little behind that of cationic liposome/micelle-based systems, even though both areas of research have been pursued for a similar length of time. However, recent results now suggest that polymer-based systems could certainly become a competitive alternative to the cationic liposome/ micelle-based systems in gene therapy research.

4.3.1 Simple polymers

The use of cationic polymers for nucleic acid delivery was pioneered with the use of poly-L-lysine (pLL) (*Figure 4.4*) which is still the most widely studied cationic polymer. This linear polymer has been produced in a variety of molecular weights (average chain lengths of 19–1116 amino acid residues [3.97–233.2 kDa]) and there is some conflict between optimal sizes for nucleic acid condensation and then delivery. Polyplexes formed at low

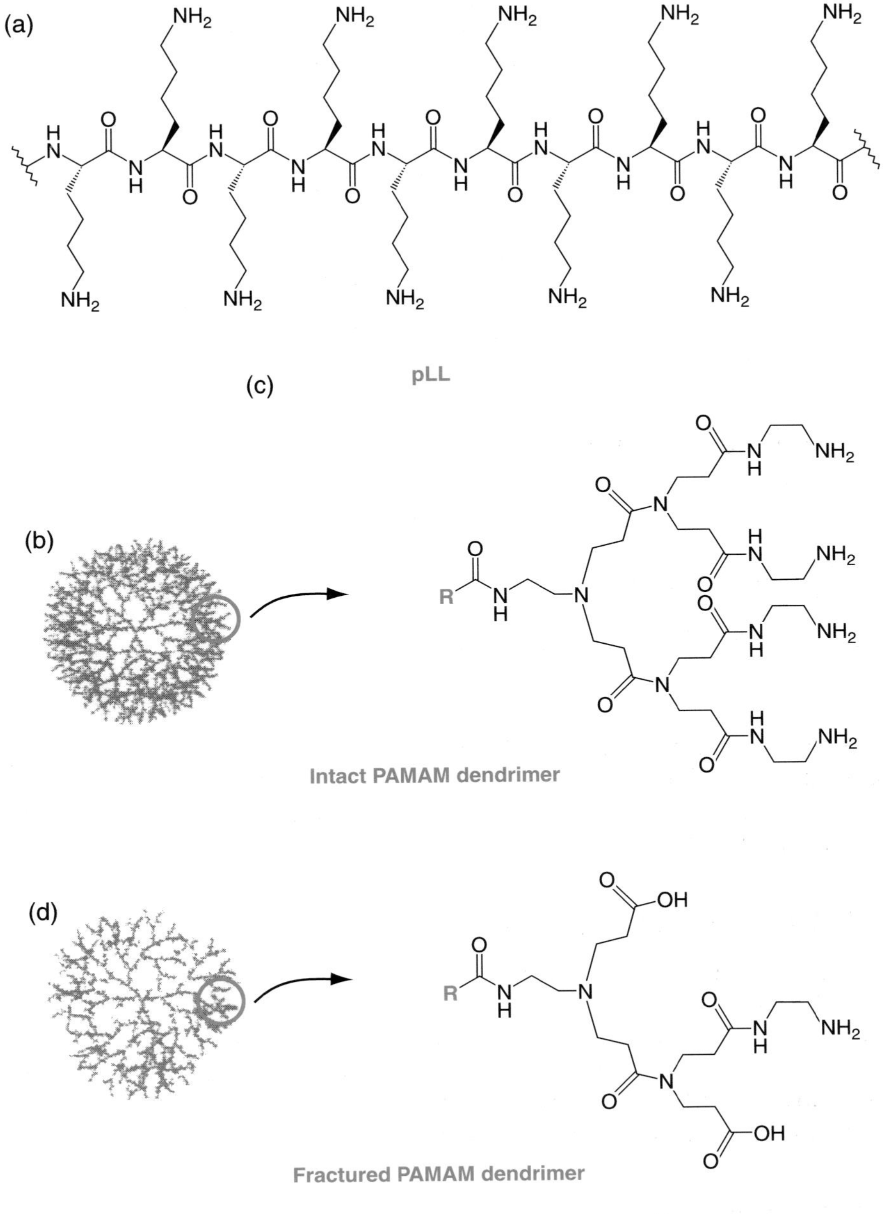

(a)

(c)

pLL

(b)

Intact PAMAM dendrimer

(d)

Fractured PAMAM dendrimer

Figure 4.4. Structures of cationic polymers used for the delivery of nucleic acids.

ionic strength with the 19 residue poly-L-lysine (pLL$_{19}$) (3.97 kDa) appear to consist of approximately spherical, monodisperse particles which vary in size from 20 to 30 nm [80]. By contrast, polyplexes formed with larger pLL polymers, such as the 114 residue pLL$_{114}$ (23.8 kDa) and above, appear to consist of particles of a larger average size (up to 300 nm) and with greater polydispersity [80]. However, at more physiological ionic strengths, this situation will partially reverse and polyplexes formed with the 1116 residue pLL$_{1116}$ (233.3 kDa) contain reasonably small, stable monodisperse particles (< 80 nm) [81]. In the case of polyplexes formed from pLL (approx 27 kDa), these have been reported to take on the appearance of toroids [82]. Nevertheless, even though well-defined polyplex particles will form, the efficiency of pLL-mediated nucleic acid delivery has been consistently found to be low in the absence of any assistance from additional agents [81–83]. For this reason, a number of research groups have explored the conjugation or incorporation of other agents either to facilitate cellular uptake *in vitro* [81, 84–88], or endosome escape [89–95]. Typically, agents chosen to facilitate cellular uptake include epidermal growth factor (EGF) peptides or transferrin; agents selected to promote endosome escape include fusogenic peptides or defective virus particles. Critically, this sort of approach has led to some successful nucleic acid delivery *in vivo* [96–98]. However, there appears to be a growing feeling that there are significant limitations to pLL-based systems which will not be easily overcome. In particular, pLL polyplexes are generally prone to aggregation [82], although attached ligands do provide some steric protection [81]. Furthermore, there is a perception that too many complicated additions need to be made to pLL-based systems in order for them to be competent agents for nucleic acid delivery [81]. Consequently, there has been a move away from pLL in recent times to find alternative cationic polymers. Before leaving this section on pLL, it is worth noting that some interesting effects have been observed using defined hexadeca-L-lysine moities attached to peptide moieties containing an arg–gly–asp (RGD) tripeptide motif. Such peptides have been found to deliver nucleic acids to cells *in vitro* by an integrin-mediated RGD-dependent cell uptake process [99–101], which mimics the way in which certain viruses such as adenovirus [102], echovirus [103] and foot-and-mouth disease viruses [104] gain entry to cells. Integrins are ubiquitous cell-surface glycoproteins involved in cell–cell and cell–matrix adhesion processes [105], many of which recognize and bind to RGD motifs in cell adhesion, serum or extracellular matrix proteins [106]. Hence, the RGD/hexadeca-L-lysine peptides are in effect copying a viral entry process and functioning as minimalist virus vectors. Doubtless, there will be much more to learn in the future from the methods employed by viruses to gain entry to cells and deliver nucleic acids to the nucleus.

Given the general interest in dendrimers in the chemistry research community, it was inevitable that some dendrimers, in particular polyamidoamine (PAMAM) dendrimers (*Figure 4.4*), would be tested as agents for nucleic acid delivery. Initially results have not been promising [107]. Intact PAMAM dendrimers appear to be too rigid to be effective agents of nucleic acid delivery [108]. Furthermore, polyplexes of PAMAM dendrimers appear to be even more prone to aggregation than their pLL counterparts [82]. By contrast, degraded or fractured PAMAM dendrimers (*Figure 4.4*) will mediate nucleic acid delivery *in vitro* with excellent efficiency and will form toroidal polyplex particles (130 $\pm$ 30 nm) stable to aggregation even in physiological buffer conditions [82, 108]. However, even more impressive than the fractured PAMAM dendrimers is polyethylenimine (PEI) (*Figure 4.5*). Like fractured PAMAM dendrimers, PEI (approx. 25 kDa; approx. 580 ethylenimine units) will also form toroidal polyplex particles (90 $\pm$ 20 nm) which are stable to aggregation in physiological buffer conditions [82], a characteristic which probably goes some way to explaining the short but impressive career of PEI as an agent for nucleic acid delivery *in vitro* and *in vivo* [109,110]. PEI also has a strong buffering capacity at almost any pH (3–10) owing to the numerous number of primary, secondary and tertiary amino groups [82]. In consequence, after a PEI polyplex has gained entry to a cell by endocytosis, nucleic acid delivery is probably also assisted by polyplex-mediated endosome buffering which it is believed will induce osmotic swelling and endosome membrane rupture allowing nucleic acids to escape into the cytosol [73]. As with pLL, a number of research groups have already begun to explore conjugation of other agents to facilitate PEI polyplex cellular uptake further *in vitro*, such as galactose, anti-CD3 antibodies, and even RGD motif-containing peptides [111–113]. Most importantly, a reasonable variety of *in vivo* applications of PEI-mediated nucleic acid delivery have now been reported, beginning with the delivery of marker genes to cells in mammalian brain [109,114]. More recently still, the slightly smaller linear PEI (ExGene 500; approx. 22 kDa; approx. 510 ethylenimine units) has proved very effective at delivering marker genes into the mouse central nervous system [115], into rabbit lung by intrapulmonary instillation [116], and into mouse lung by systemic administration routes [117]. There has also even been a recent report of galactosyl-PEI being used successfully *in vivo* [118]. However, a word of caution needs to be inserted here. In contrast to cationic liposome/micelle-based systems, PEI has been reported to show quite serious toxic effects *in vivo* even though the smaller linear 22 kDa PEI appears to have a better toxicity profile than the original 25 kDa PEI [117]. Unfortunately, this may mean that PEI will only have a limited application in the future, in spite of impressive results obtained so far.

4.3.2 Alternative polymer systems

A number of alternative polymer systems have been reported to mediate nucleic acid delivery. For instance, there is the cationic polymer APL

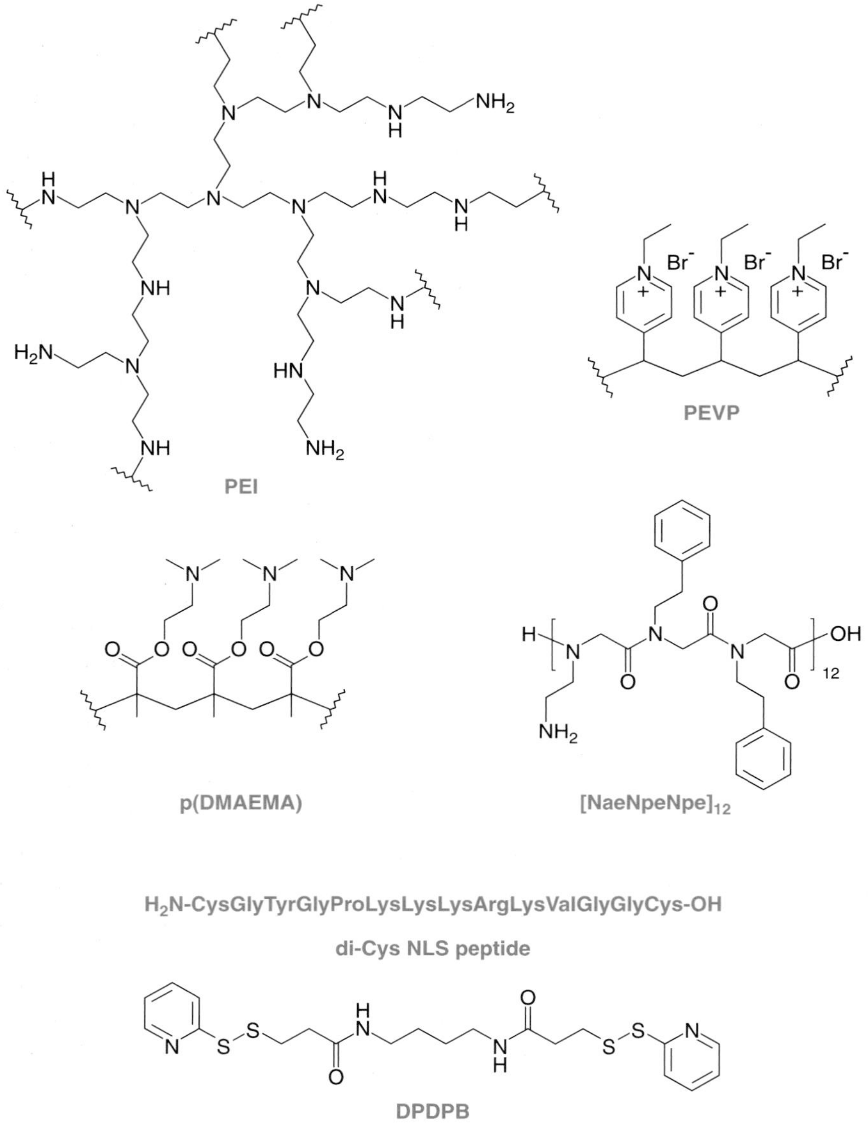

PEI

PEVP

p(DMAEMA)

[NaeNpeNpe]₁₂

H₂N-CysGlyTyrGlyProLysLysLysArgLysValGlyGlyCys-OH

di-Cys NLS peptide

DPDPB

Figure 4.5. Structures of cationic polymers used for the delivery of nucleic acids.

PolyCat57 which has been described as a glucaramide-based nonpeptide, nonlipid polymer with a charge profile that predicts its spontaneous interaction with plasmid DNA [119]. Polyplexes of this polymer have been reported to be resistant to serum inhibition and efficient at gene delivery *in vivo* following direct injection into intracranial human glioma cell xenografts in mice brains [119]. Another class of synthetic polymers which has been reported to deliver nucleic acids into mammalian cells *in vitro* is represented by poly(*N*-ethyl-4-vinylpyridinium bromide) (PEVP) (*Figure 4.5*) [120]. However, polyplexes of this polymer appear to be unstable to aggregation and seem to require the presence of micelles of Pluronic acid P85 [a poly(ethylene oxide)-*block*-poly(propylene oxide)-*block*-poly(ethylene oxide) copolymer] to promote nucleic acid delivery in preference to aggregation [121]. Poly(2-(dimethylamino)ethyl methacrylate) [p(DMAE-MA)] (*Figure 4.5*) type polymers have also been evaluated as agents for nucleic acid delivery *in vitro* although in this instance cytotoxicity problems have cast some doubt over the long term future of such polymeric systems in gene therapy research [122–124]. Block copolymers containing a number of the above polymeric elements have also been examined [83]. Turning to other alternative polymer systems, cationic histones [125] and HMG proteins [126] have proved useful for *in vitro* nucleic acid delivery, not to mention a number of short-peptide systems, some with a propensity to form amphiphilic α-helices, which have all shown some facility to deliver nucleic acids *in vitro* [127–130]. An intriguing variation to this theme has been the use of peptoid polymers, such as [NaeNpeNpe]$_{12}$ (*Figure 4.5*). This was generated by a solid-phase combinatorial chemistry procedure [131], based upon some high yielding chemistry previously reported [132]. This polymer was found to be at least as effective as some cationic liposome systems at mediating nucleic acid delivery *in vitro* although there have been no reports of use *in vivo* [131]. Finally, one of the most imaginative recent approaches to the generation of novel polymer systems has been to regard nucleic acids, such as DNA, as a template for polymer generation [133]. In so doing, the nascent polymer is proposed to optimally condense DNA into polyplex particles which are able to mediate nucleic acid delivery. The best reported example of this approach so far has been to condense plasmid DNA using a template polymerization mixture containing a 14-mer peptide encoding the nuclear localization sequence (NLS) of SV40 T antigen (di-Cys-NLS) with a cross-linking agent 1,4-di[3′,2′-pyridyldithio(propionamido)butane] (DPDPB) (*Figure 4.5*). The resulting polyplex mixture was able to mediate plasmid delivery *in vitro* though once again there have been no reports of use *in vivo* [133].

4.4 Alternative chemical nonviral delivery systems

Bearing in mind the failure of either cationic liposome/micelle or polymer-based systems to deliver nucleic acids to the level required for clinical gene therapy applications, there have been some recent attempts to develop completely alternative systems based upon highly selective nucleic acid binding systems. For instance, a number of reports describing the use of chimeric proteins which contain both cell targeting, translocation and DNA-binding domains have been made [134,135] (*Figure 4.6*). These constructs actually have an overall negative charge and must be used in

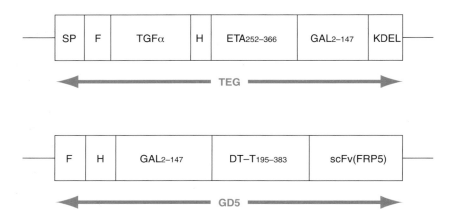

Figure 4.6. Diagrammatic representation of two chimeric fusion protein nonviral delivery systems. The chimeric **TEG** protein consists of an N-terminal *Escherichia coli ompA* signal peptide (**SP**), the synthetic FLAG epitope (**F**), amino acid residues 1–50 of human **TGFα**, a polyhistidine tag (**H**), amino acid residues 252–366 of Pseudomonas exotoxin A (translocation domain) (**ETA**), amino acid residues 2–147 of the yeast **GAL4** (DNA-binding domain), and a C-terminal **KDEL** endoplasmic reticulum retention signal. The chimeric **GD5** protein consists of some identical elements to the TEG protein, namely an *N*-terminal **F**, **H** and **GAL4**, but followed by amino acid residues 195–383 of the bacterial diphtheria toxin (translocation domain) (**DT-T**) and a *C*-terminal ErbB2-specific **scFv(FRP5)** antibody domain.

conjunction with pLL if they are actually to be able to deliver nucleic acids properly to cells. This creates some limitations and the gene delivery efficiencies of the chimeric protein/pLL polyplexes are yet only modest. However, this may point to nonviral delivery systems in the future which only contain closely defined nucleic acid binding regions complementary to specific nucleic acid base sequences, and attached ligands for receptor mediated uptake. Such systems could then dispense with the need for cationic entities to condense nucleic acids, instead causing the effective transport of naked nucleic acids into cells. In such an event, the nucleic acids themselves could be modified by the attachment of nuclear localization sequences in order to assist the delivered nucleic acids enter the nucleus [136,137]. Certainly, there is growing belief amongst some researchers in the field of chemical nonviral delivery systems that this could represent a potent future for nonviral vector systems.

4.5 Physical nonviral delivery systems

Whilst the majority of this chapter has been concerned with chemical nonviral delivery systems for reasons explained in the introduction, there are a number of physical nonviral delivery methods under development.

Broadly speaking, the physical methods are electroporation, micro-injection and biolistics. Of these, electroporation and biolistics are showing real promise as methods for the delivery of nucleic acids *in vivo*. Therefore, I shall confine my remarks to these two techniques. Biolistics was first reported a number of years ago as a means of physically projecting molecules into cells or tissues of interest [138]. Since that time, the technique has been quickly adapted for the delivery of nucleic acids into a variety of cells and tissues *in vitro* and *in vivo* [139–142]. A current gene gun (Helios; Bio-Rad) developed for the purpose of biolistic delivery of nucleic acids uses a high velocity jet of helium (900–1800 psi) to collisionally accelerate nucleic acid-coated metal (gold or tungsten; 0.8–1.2 µm) microcarriers. As a result, the microcarriers become high velocity microprojectiles capable of entering the cytoplasm of target cells and in so doing delivering the nucleic acids coated on their surface (*Figure 4.7*). This technique could be extended to further *in vivo* applications, most especially where exposed epithelial surfaces are concerned. However, such a physical technique would always carry an attendant risk of cell damage which could only be minimized by a very extensive analysis of all the physical parameters and variables involved in this particular nonviral delivery system. Nethertheless, the real possibility exists that this technique could be a useful tool for future gene therapy research. In a similar way, electroporation appears to be very promising, promoting quite impressive levels of nucleic acid delivery to a variety of tissues *in vivo* including skin, muscle, brain and liver [143,144]. Whilst this technique has been familiar for many years as a routine laboratory method for the transfection of cultured mammalian or bacterial cells, recent research has shown how clusters of electrodes inserted into tissue will emit controlled electrical pulses able to create transmembrane potentials sufficient to allow locally administered naked DNA to enter cells, probably through transient pores [145]. Local tissue damage will result, but, fortunately, tissue regeneration appears to be rapid [143]. The general utility of electroporation in gene therapy research remains to be established.

4.6 Future prospects for nonviral delivery systems

The process of developing successful nonviral delivery systems for use *in vivo* is still really in its infancy. A range of different nonviral delivery systems have been devized but no one system is near well developed enough for use in clinical gene therapy with the possible exception of some cationic liposome/micelle-based systems. Therefore, for the most part the different nonviral delivery systems currently available appear to represent useful starting points in the quest for clinically useful nonviral delivery systems but are not more significant than that as yet. Quite likely, the final clinical nonviral delivery systems of choice will contain elements of these starting points but in most respects will probably look very different.

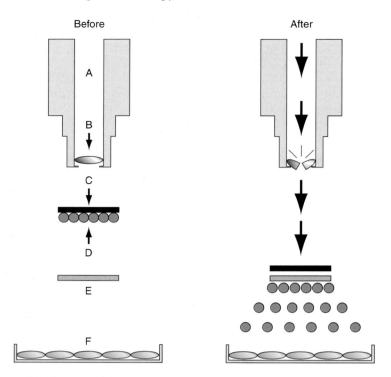

Figure 4.7. Diagrammatic representation of a gene gun. Before activation, the labels are as follows; **A**: gas acceleration tube; **B**: rupture disc; **C**: macrocarrier; **D**: microcarriers coated with nucleic acids; **E**: stopping screen; **F**: target cells. After activation, helium gas (travelling in the direction of the large arrows) propels the macrocarrier as far as the stopping screen whereupon the microcarriers are released with sufficient momentum to penetrate the cell membranes of the target cells and deliver the nucleic acids coated on their surfaces.

However, given the relative youth of this field and the increasing move to systematic studies designed to tackle problems of the type outlined earlier (*Section 4.1.2*), there is no reason to believe that nonviral delivery systems equal to the task of human gene therapy cannot be developed within the foreseeable future.

References

1. Felgner, P.L., Barenholz,Y., Behr, J.-P., Cheng, S.H., Cullis, P., Huang, L., Jessee, J.A., Seymour, L., Szoka, F., Thierry, A.R., Wagner, E. and Wu, G. (1997) Nomenclature for synthetic gene delivery systems. *Hum. Gene Ther.*, **8**, 511–512.
2. Miller, A.D. (1998) Cationic liposomes for gene therapy. *Angew. Chem. Intl. Ed.*, **37**, 1768–1785.
3. Miller, A.D. (1998) Cationic liposome systems in gene therapy. *Curr. Res. Mol. Ther.*, **1**, 494–503.

4. Felgner, P.L., Gadek, T.R., Holm, M., Roman R., Chan H.W., Wenz, M., Northrop, J.P., Ringold, G.M. and Danielsen, M. (1987) Lipofection: a highly efficient, lipid-mediated DNA-transfection procedure. *Proc. Natl Acad. Sci. USA*, **84**, 7413–7417.

5. Leventis, R. and Silvius, J.R. (1990) Interactions of mammalian cells with lipid dispersions containing novel metabolizable cationic amphiphiles. *Biochim. Biophys. Acta*, **1023**, 124–132.

6. Gebeyehu, G., Jessee, J.A., Valentina, C. and Hawley-Nelson, P. (1993) Cationic lipids. US-05334761, GIBCO/BRL.

7. Nantz, M.H., Bennett, M.J. and Malone, R.W. (1996) Cationic transport reagents. US-05527928, Promega.

8. Behr, J.-P., Demeneix, B., Loeffler, J.-P. and Perez-Mutul, J. (1989) Efficient gene transfer into mammalian primary endocrine cells with lipopolyamine-coated DNA. *Proc. Natl Acad. Sci. USA*, **86**, 6982–6986.

9. Ruysschaert, J.-M., El Ouahabi A., Willeaume, V., Huez, G., Fuks, R., Vandenbranden, M. and Di Stefano, P. (1994) A novel cationic amphiphile for transfection of mammalian cells. *Biochem. Biophys. Res. Commun.*, **203**, 1622–1628.

10. Defrise-Quertain, F., Duquenoy, P., Brasseur, R., Brak, P., Caillaux, B., Fuks, R. and Ruysschaert, J.-M. (1986) Vesicle formation by double long-chain amidines. *J. Chem. Soc. Chem. Commun.*, 1060–1062.

11. Kunitake, T., Okahata, Y., Tamaki, K., Kumamaru, F. and Takayanagi, M. (1977) Formation of the bilayer membrane from a series of quaternary ammonium salts. *Chem. Lett.*, 387–390.

12. Moradpour, D., Schauer, J.I., Zurawski, V.R. Jr., Wands, J.R. and Boutin, R.H. (1996) Efficient gene transfer into mammalian cells with cholesterol-spermidine. *Biochem. Biophys. Res. Commun.*, **221**, 82–88.

13. Gao, X. and Huang, L. (1991) A novel cationic liposome reagent for efficient transfection of mammalian cells. *Biochem. Biophys. Res. Commun.*, **179**, 280–285.

14. Felgner, J.H., Kumar, R., Sridhar, C.N., Wheeler, C.J., Tsai, Y.J., Border, R., Ramsey, P., Martin, M. and Felgner, P.L. (1994) Enhanced gene delivery and mechanism studies with a novel series of cationic lipid formulations. *J. Biol. Chem.*, **269**, 2550–2561.

15. Wheeler, C.J., Felgner, P.L., Tsai Y.J., Marshall, J., Sukhu, L., Doh, S.G., Hartikka, J., Nietupski, J., Manthorpe, M., Nichols, M., Plewe, M., Liang, X., Norman, J., Smith, A. and Cheng, S. H. (1996) A novel cationic lipid greatly enhances plasmid DNA delivery and expression in mouse lung. *Proc. Natl Acad. Sci. USA*, **93**, 11454–11459.

16. Balasubramaniam, R.P., Bennett, M.J., Aberle, A.M., Malone, J.G., Nantz, M.H. and Malone, R.W. (1996) Structural and functional analysis of cationic transfection lipids: the hydrophobic domain. *Gene Ther.*, **3**, 163–172.

17. Akao, T., Nakayama, T., Takeshia, K. and Ito, A. (1994) Design of a new cationic amphiphile with efficient DNA-transfection ability. *Biochem. Mol. Biol. Intl.*, **34**, 915–920.

18. Kunitake, T., Nakashima, N., Shimomura, M., Okahata, Y., Kano, K. and Ogawa, T. (1980) Unique properties of chromophore-containing bilayer aggregates: enhanced chirality and photo-chemically induced morphological change. *J. Am. Chem. Soc.*, **102**, 6642–6644.

19. Lewis, J.G., Lin, K.-Y., Kothavale, A., Flanagan, W.M., Matteucci, M.D., DePrince, B., Mook, R.A. Jr., Hendren R.W. and Wagner R.W. (1996) A serum-resistant cytofectin for cellular delivery of antisense oligodeoxynucleotides and plasmid DNA. *Proc. Natl Acad. Sci. USA*, **93**, 3176–3181.

20. Pitard, B., Aguerre, O., Airiau, M., Lachages, A.-M., Boukhnikachvili, T., Byk, G., Dubertret, C., Herviou, C., Scherman, D., Mayaux, J.-F. and Crouzet, J. (1997) Virus-sized self assembling lamellar complexes between plasmid DNA and cationic micelles promote gene transfer. *Proc. Natl Acad. Sci. USA*, **94**, 14412–14417.

21. Byk, G., Dubertret, C., Escriou, V., Frederic, M., Jaslin, G., Rangara, R., Pitard, B., Crouzet, J., Wils, P., Schwartz, B. and Scherman, D. (1998) Synthesis, activity and structure-activity relationship studies of novel cationic lipids for DNA transfer. *J. Med. Chem.*, **41**, 224–235.

22. Thierry, A.R., Rabinovich, P., Peng, B., Mahan, L.C., Bryant, J.L. and Gallo, R.C. (1997) Characterization of liposome-mediated gene delivery: expression, stability and pharmacokinetics of plasmid DNA. *Gene Ther.*, **4**, 226–237.

23. Walker, S., Sofia, M.J., Kakarla, R., Kogan N.A., Wierichs, L., Longley, C.B., Bruker, K., Axelrod, H.R., Midha, S., Babu, S. and Kahne, D. (1996) Cationic facial amphiphiles: a promising class of transfection agents. *Proc. Natl Acad. Sci. USA*, **93**, 1585–1590.

24. Lee, E.R., Marshall, J., Siegel, C.S., Jiang, C., Yew, N.S., Nichols M.R., Nietupski, J.B., Ziegler, R.J., Lane, M.B., Wang, K.X., Wan, N.C., Scheule, R.K., Harris, D.J., Smith, A.E. and Cheng, S.H. (1996) Detailed analysis of structures and formulations of cationic lipids for efficient gene transfer to the lung. *Hum. Gene Ther.*, **7**, 1701–1717.

25. Cooper, R.G., Etheridge, C.J., Stewart, L., Marshall, J., Rudginsky, S., Cheng, S.H. and Miller, A.D. (1998) Polyamine analogues of 3β-[N-(N',N'-dimethylaminoethane)carbamoyl]cholesterol (DC-Chol) as agents for gene delivery. *Chem. Eur. J.*, **4**, 137–152.

26. Vigneron, J.P., Oudrhiri, N., Fauquet, M., Vergely, L., Bradley, J.-C., Basseville, M., Lehn, P. and Lehn, J.-M. (1996) Guanidinium-cholesterol cationic lipids: efficient vectors for the transfection of eukaryotic cells. *Proc. Natl Acad. Sci. USA*, **93**, 9682–9686.

27. Puyal, C., Milhaud, P., Bienvenüe, A. and Philippot, J.R. (1995) A new cationic liposome encapsulating genetic material. A potential delivery system for polynucleotides. *Eur. J. Biochem.*, **228**, 697–703.

28. Brunette, E., Stribling, R. and Debs, R. (1992) Lipofection does not require the removal of serum. *Nucleic Acids Res.*, **20**, 1151.

29. Gorman, C.M., Aikawa, M., Fox, B., Fox, E., Lapuz, C., Michaud, B., Nguyen, H., Roche, E., Sawa, T. and Wiener-Kronish, J.P. (1997) Efficient *in vivo* delivery of DNA to pulmonary cells using the novel lipid EDMPC. *Gene Ther.*, **4**, 983–992.

30. Ito, A., Miyazoe, R., Mitoma, J.-y., Akao, T., Osaki, T. and Kunitake, T. (1990) Synthetic cationic amphiphiles for liposome-mediated DNA transfection. *Biochem. Intl.*, **22**, 235–241.

31. Solodin, I., Brown, C.S., Bruno, M.S., Chow, C.-Y., Jang, E.-H., Debs, R.J. and Heath, T.D. (1995) A novel series of amphiphilic imidazolinium componds for *in vitro* and *in vivo* gene delivery. *Biochemistry*, **34**, 13537–13544.

32. van der Woude, I., Wagenaar, A., Meekel, A.A.P., TerBeest, M.B.A., Ruiters, M.H.J., Engberts, J.B.F.N. and Hoekstra, D. (1997) Novel pyridinium surfactants for efficient, nontoxic *in vitro* gene delivery. *Proc. Natl Acad. Sci. USA*, **94**, 1160–1165.

33. Dodds, E., Dunckley, M.G., Naujoks, K., Michaelis, U. and Dickson, G. (1998) Lipofection of cultured mouse muscle cells: a direct comparison of Lipofectamine and DOSPER. *Gene Ther.*, **5**, 542–551.

34. Vitiello, L., Bockhold, K., Joshi, P.B. and Worton, R.G. (1998) Transfection of cultured myoblasts in high serum concentration with DODAC:DOPE liposomes. *Gene Ther.*, **5**, 1306–1313.
35. Schwartz, B., Benoist, C., Abdallah, B., Scherman, D., Behr, J.-P. and Demeneix, B.A. (1995) Lipospermine-based gene-transfer into the newborn mouse-brain is optimized by a low lipospermine DNA charge ratio. *Hum. Gene Ther.*, **6**, 1515–1524.
36. Liu, F., Qi, H., Huang, L. and Liu, D. (1997) Factors controlling the efficiency of cationic lipid-mediated transfection *in vivo* via intravenous administration. *Gene Ther.*, **4**, 517–523.
37. Yang, J.-P. and Huang, L. (1997) Overcoming the inhibitory effect of serum on lipofection by increasing the charge ratio of cationic liposome to DNA. *Gene Ther.*, **4**, 950–960.
38. Remy, J.-S., Kichler, A., Mordvinov, V., Schuber, F. and Behr, J.-P. (1995) Targeted gene transfer into hepatoma cells with lipopolyamine-condensed DNA particles presenting galactose ligands: a stage toward artificial viruses. *Proc. Natl Acad. Sci. USA*, **92**, 1744–1748.
39. Alton, E.W.F.W., Middleton, P.G., Caplen, N.J., Smith, S.N., Steel, D.M., Munkonge, F.M., Jeffery, P.K., Geddes, D.M., Hart, S.L., Williamson, R., Fasold, K.I., Miller, A.D., Dickinson, P., Stevenson, B.J., McLachlan, G., Dorin, J.R. and Porteous, D.J. (1993) Non-invasive liposome-mediated gene delivery can correct the ion transport defect in cystic fibrosis mutant mice. *Nat. Genet.*, **5**, 135–142.
40. McQuillin, A., Murray, K.D., Etheridge, C.J., Stewart, L., Cooper, R.G., Brett, P.M., Miller, A.D. and Gurling, H.M.D. (1997) Optimization of liposome mediated transfection of a neuronal cell line. *NeuroReport*, **8**, 1481–1484.
41. Fife, K., Bower, M., Cooper, R.G., Stewart, L., Etheridge, C.J., Coombes, R.C., Buluwela, L. and Miller, A.D. (1998) Endothelial cell transfection with cationic liposomes and herpes simplex-thymidine kinase mediated killing. *Gene Ther.*, **5**, 614–620.
42. Labat-Moleur, F., Steffan, A.-M., Brisson, C., Perron, H., Feugeas, O., Furstenberger, P., Oberling, F., Brambilla, E. and Behr, J.-P. (1996) An electron microscopy study into the mechanism of gene transfer with lipopolyamines. *Gene Ther.*, **3**, 1010–1017.
43. Zabner, J., Fasbender, A.J., Moninger, T., Poellinger, K.A. and Welsh, M.J. (1995) Cellular and molecular barriers to gene transfer by a cationic lipid. *J. Biol. Chem.*, **270**, 18997–19007.
44. Rädler, J.O., Koltover, I., Salditt, T. and Safinya, C.R. (1997) Structure of DNA-cationic liposome complexes: DNA intercalation in multilamellar membranes in distinct interhelical packing regimes. *Science,* **275**, 810–814.
45. Gustafsson, J., Arvidson, G., Karlsson, G. and Almgren, M. (1995) Complexes between cationic liposomes and DNA visualized by cryo-TEM. *Biochim. Biophys. Acta*, **1235**, 305–312.
46. Gershon, H., Ghirlando, R., Guttman, S.B. and Minsky, A. (1993) Mode of formation and structural features of DNA-cationic liposome complexes used for transfection. *Biochemistry*, **32**, 7143–7151.
47. Eastman, S.J., Siegel, C., Tousignant, J., Smith, A.E., Cheng, S.H. and Scheule, R.K. (1997) Biophysical characterisation of cationic lipid:DNA complexes. *Biochim. Biophys Acta*, **1325**, 41–62.
48. Sternberg, B., Sorgi, F.L. and Huang, L. (1994) New structures in complex formation between DNA and cationic liposomes visualised by freeze-fracture electron microscopy. *FEBS Lett.*, **356**, 361–366.

49. Stribling, R., Brunette, E., Liggitt, D., Gaensler, K. and Debs, R. (1992) Aerosol gene delivery *in vivo*. *Proc. Natl Acad. Sci. USA*, **89**, 11277–11281.

50. Caplen, N.J., Alton, E.W.F.W., Middleton, P.G., Dorin, J.R., Stevenson, B.J., Gao, X., Durham, S.R., Jeffery, P.K., Hodson, M.E., Coutelle, C., Huang, L., Porteous, D.J., Williamson, R. and Geddes, D.M. (1995) Liposome-mediated CFTR gene transfer to the nasal epithelium of patients with cystic fibrosis. *Nat. Med.*, **1**, 39–46.

51. Alton, E.W.F.W., Stern, M., Farley, R. *et al.* (1999) Cationic lipid-mediated CFTR gene transfer to the lungs and nose of patients with cystic fibrosis: a double-blind placebo-controlled trial. *Lancet*, **353**, 947–954.

52. Zhu, N., Liggitt, D., Liu, Y. and Debs, R. (1993) Systemic gene-expression after intravenous DNA delivery into adult mice. *Science*, **261**, 209–211.

53. Rogy, M.A., Auffenberg, T., Espat, N.J., Philip, R., Remick, D., Wollenberg, G.K., Copeland III, E.M. and Moldawer, L.L. (1995) Human tumor necrosis factor receptor (p55) and interleukin 10 gene transfer in the mouse reduces mortality to lethal endotoxemia and also attenuates local inflammatory responses. *J. Exp. Med.*, **181**, 2289–2293.

54. Liu, Y., Liggitt, D., Zhong, W., Tu, G., Gaensler, K. and Debs, R. (1995) Cationic liposome-mediated intravenous gene delivery. *J. Biol. Chem.*, **270**, 24864–24870.

55. Thierry, A.R., Lunardiiskandar, Y., Bryant, J.L., Rabinovich, P., Gallo, R.C. and Mahan, L.C. (1995) Systemic gene therapy: biodistribution and long-term expression of a transgene in mice. *Proc. Natl Acad. Sci. USA*, **92**, 9742–9746.

56. Murray, K.D., McQuillin, A., Stewart, L., Etheridge, C.J., Cooper, R.G., Miller, A.D. and Gurling, H.M.D. (1999) Cationic liposome-mediated DNA transfection in organotypic explant cultures of the ventral mesencephalon. *Gene Ther.*, **6**, 190–197.

57. Stephan, D.J., Yang, Z.-Y., San, H., Simari, R.D., Wheeler, C.J., Felgner, P.L., Gordon, D., Nabel, G.J. and Nabel, E.G. (1996) A new cationic liposome DNA complex enhances the efficiency of arterial gene transfer *in vivo*. *Hum. Gene Ther.*, **7**, 1803–1812.

58. Seung, L.P., Mauceri, H.J., Beckett, M.A., Hallahan, D.E., Hellman, S. and Weichselbaum, R.R. (1995) Genetic radiotherapy overcomes tumor resistance to cytotoxic agents. *Cancer Res.*, **55**, 5561–5565.

59. Zhu, J., Zhang, L., Hanisch, U.K., Felgner, P.L. and Reszka, R. (1996) A continuous intracerebral gene delivery system for *in vivo* liposome-mediated gene therapy. *Gene Ther.*, **3**, 472–476.

60. Wright, M.J., Rosenthal, E., Stewart, L., Wightman, L.M.L., Miller A.D., Latchman, D.S. and Marber, M.S. (1998) β-Galactosidase staining following intracoronary infusion of cationic liposomes in the *in vivo* rabbit heart is produced by microinfarction rather than effective gene transfer: a cautionary tale. *Gene Ther.*, **5**, 301–308.

61. Hong, K., Zheng, W., Baker, A. and Papahadjopoulos, D. (1997) Stabilization of cationic liposome-plasmid DNA complexes by polyamines and poly(ethylene glycol)-phospholipid conjugates for efficient *in vivo* gene delivery. *FEBS Lett.*, **400**, 233–237.

62. Hofland, H.E.J., Shephard, L. and Sullivan, S.M. (1996) Formation of stable cationic lipid/DNA complexes for gene transfer. *Proc. Natl Acad. Sci. USA*, **93**, 7305–7309.

63. Litzinger, D.C., Brown, J.M., Wala, I., Kaufman, S.A., Van, G.Y., Farrell, C.L. and Collins, D. (1996) Fate of cationic liposomes and their complex with oligonucleotide *in vivo*. *Biochim. Biophys. Acta*, **1281**, 139–149.

64. Plank, C., Mechtler, K., Szoka, F.C. Jr. and Wagner, E. (1996) Activation of the complement system by synthetic DNA complexes: a potential barrier for intravenous gene delivery. *Hum. Gene Ther.*, **7**, 1437–1446.
65. Yang, J.-P. and Huang, L. (1998) Time-dependent maturation of cationic liposome-DNA complex for serum resistance. *Gene Ther.*, **5**, 380–387.
66. Liu, Y., Mounkes, L.C., Liggitt, H.D., Brown, C.S., Solodin, I., Heath, T.D. and Debs, R.J. (1997) Factors influencing the efficiency of cationic liposome-mediated intravenous gene delivery. *Nat. Biotech.*, **15**, 167–173.
67. Templeton, N.S., Lasic, D.D., Frederik, P.M., Strey, H.H., Roberts, D.D. and Pavlakis, G.N. (1997) Improved DNA:liposome complexes for increased systemic delivery and gene expression. *Nat. Biotech.*, **15**, 647–652.
68. Spragg, D.D., Alford, D.R., Greferath, R., Larsen, C.E., Lee, K.-D., Gurtner, G.C., Cybulsky, M.I., Tosi, P.F., Nicolau, C. and Gimbrone, M.A. Jr. (1997) Immunotargeting of liposomes to activated vascular endothelial cells: a strategy for site-directed delivery in the cardiovascular system. *Proc. Natl Acad. Sci. USA*, **94**, 8795–8800.
69. Simeõs, S., Slepushkin, V., Gaspar, R., Pedroso de Lima, M.C. and Düzgünes, N. (1998) Gene delivery by negatively charged ternary complexes of DNA, cationic liposomes and transferrin or fusigenic peptides. *Gene Ther.*, **5**, 955–964.
70. Kamata, H., Yagisawa, H., Takahashi, S. and Hirata, H. (1994) Amphiphilic peptides enhance the efficiency of liposome-mediated DNA transfection. *Nucleic Acids Res.*, **22**, 536–537.
71. Budker, V., Gurevich, V., Hagstrom, J.E., Bortzov, F. and Wolff, J.A. (1996) pH-sensitive, cationic liposomes: a new synthetic virus-like vector. *Nat. Biotech.*, **14**, 760–764.
72. Legendre, J.Y. and Szoka, F.C. (1992) Delivery of plasmid DNA into mammalian-cell lines using pH-sensitive liposomes: comparison with cationic liposomes. *Pharmaceut. Res.*, **9**, 1235–1242.
73. Behr, J.-P. (1996) The proton sponge, a means to enter cells viruses never thought of. *M/S-Med. Sci.*, **12**, 56–58.
74. Li, S. and Huang, L. (1997) *In vivo* gene transfer via intravenous administration of cationic lipid-protamine-DNA (LPD) complexes. *Gene Ther.*, **4**, 891–900.
75. Li, S., Rizzo, M.A., Bhattacharya, S. and Huang, L. (1998) Characterization of cationic lipid-protamine-DNA (LPD) complexes for intravenous gene delivery. *Gene Ther.*, **5**, 930–937.
76. Yonemitsu, Y., Alton, E.W.F.W., Komori, K., Yoshizumi, T., Sugimachi, K. and Kaneda, Y. (1998) HVJ (Sendai virus) liposome-mediated gene transfer: current status and future perspectives. *Int. J. Oncol.*, **12**, 1277–1285.
77. Kaneda, Y., Iwai, K. and Uchida, T. (1989) Increased expression of DNA cointroduced with nuclear-protein in adult-rat liver. *Science*, **243**, 375–378.
78. Saeki, Y., Matsumoto, N., Nakano, Y., Mori, M., Awai, K. and Kaneda, Y. (1997) Development and characterization of cationic liposomes conjugated with HVJ (Sendai virus): reciprocal effect of cationic lipid for *in vitro* and *in vivo* gene transfer. *Hum. Gene Ther.*, **8**, 2133–2141.
79. Yonemitsu, Y., Kaneda, Y., Muraishi, A., Yoshizumi, T., Sugimachi, K. and Sueishi, K. (1997) HVJ (Sendai virus)-cationic liposomes: a novel and potentially effective liposome-mediated technique for gene transfer to the airway epithelium. *Gene Ther.*, **4**, 631–638.
80. Wolfert, M.A. and Seymour, L.W. (1996) Atomic force microscopic analysis of the influence of the molecular weight of poly(L)lysine on the size of polyelectrolyte complexes formed with DNA. *Gene Ther.*, **3**, 269–273.

81. Xu, B., Wiehle, S., Roth, J.A. and Cristiano, R.J. (1998) The contribution of poly-L-lysine, epidermal growth factor and streptavidin to EGF/PLL/DNA polyplex formation. *Gene Ther.*, **5**, 1235–1243.

82. Tang, M.X. and Szoka, F.C. (1997) The influence of polymer structure on the interactions of cationic polymers with DNA and morphology of the resulting complexes. *Gene Ther.*, **4**, 823–832.

83. Wolfert, M.A., Schacht, E.H., Toncheva, V., Ulbrich, K., Nazarova, O. and Seymour, L.W. (1996) Characterization of vectors for gene therapy formed by self-assembly of DNA with synthetic block co-polymers. *Hum. Gene Ther.*, **7**, 2123–2133.

84. Wu, G.Y. and Wu, C. H. (1988) Evidence for targeted gene delivery to HepG2 hepatoma-cells *in vitro*. *Biochemistry*, **27**, 887–892.

85. Michael, S.I. and Curiel, D.T. (1994) Strategies to achieve targeted gene delivery via the receptor-mediated endocytosis pathway. *Gene Ther.*, **1**, 223–232.

86. Findeis, M.A., Wu, C.H. and Wu, G.Y. (1994) Ligand-based carrier systems for delivery of DNA to hepatocytes. *Meth. Enzymol.*, **247**, 341–351.

87. Kollen, W.J.W., Midoux, P., Erbacher, P., Yip, A., Roche, A.C., Monsigny, M., Glick, M.C. and Scanlin, T.F. (1996) Gluconoylated and glycosylated polylysines as vectors for gene transfer into cystic fibrosis airway epithelial cells. *Hum. Gene Ther.*, **7**, 1577–1586.

88. Erbacher, P., Bousser, M.-T., Raimond, J., Monsigny, M., Midoux, P. and Roche, A.C. (1996) Gene transfer by DNA/glycosylated polylysine complexes into human blood monocyte-derived macrophages. *Hum. Gene Ther.*, **7**, 721–729.

89. Cotten, M., Langlerouault, F., Kirlappos, H., Wagner, E., Mechtler, K., Zenke, M., Beug, H. and Birnstiel, M.L. (1990) Transferrin polycation-mediated introduction of DNA into human leukemic cells: stimulation by agents that affect the survival of transfected DNA or modulate transferrin receptor levels. *Proc. Natl Acad. Sci. USA*, **87**, 4033–4037.

90. Plank, C., Zatloukal, K., Cotten, M., Mechtler, K. and Wagner, E. (1992) Gene transfer into hepatocytes using asialoglycoprotein receptor mediated endocytosis of DNA complexed with an artificial tetra-antennary galactose ligand. *Bioconjugate Chem.*, **3**, 533–539.

91. Midoux, P., Mendes, C., Legrand, A., Raimond, J., Mayer, R., Monsigny, M. and Roche, A.C. (1993) Specific gene transfer mediated by lactosylated poly-L-lysine into hepatoma-cells. *Nucleic Acids Res.*, **21**, 871–878.

92. Erbacher, P., Roche, A.C., Monsigny, M. and Midoux, P. (1996) Putative role of chloroquine in gene transfer into a human hepatoma cell line by DNA lactosylated polylysine complexes. *Exp. Cell Res.*, **225**, 186–194.

93. Plank, C., Oberhauser, B., Mechtler, K., Koch, C. and Wagner, E. (1994) The influence of endosome-disruptive peptides on gene transfer using synthetic virus-like gene transfer systems. *J. Biol. Chem.*, **17**, 12918–12924.

94. Curiel, D.T., Agarwal, S., Wagner, E. and Cotten, M. (1991) Adenovirus enhancement of transferrin polylysine-mediated gene delivery. *Proc. Natl Acad. Sci. USA*, **88**, 8850–8854.

95. Wagner, E., Zatloukal, K., Cotten, M., Kirlappos, H., Mechtler, K., Curiel, D.T. and Birnstiel, M.L. (1992) Coupling of adenovirus to transferrin-polylysine/DNA complexes greatly enhances receptor-mediated gene delivery and expression of transfected genes. *Proc. Natl Acad. Sci. USA*, **89**, 6099–6103.

96. Wu, G.Y., Wilson, J.M., Shalaby, F., Grossman, M., Shafritz, D.A. and Wu, C.H. (1991) Receptor-mediated gene delivery *in vivo*-partial correction of genetic analbuminemia in nagase rats. *J. Biol. Chem.*, **266**, 14338–14342.

97. Perales, J.C., Ferkol, T., Molas, M. and Hanson, R.W. (1994) An evaluation of receptor-mediated gene-transfer using synthetic DNA-ligand complexes. *Eur. J. Biochem.*, **226**, 255–266.
98. Ferkol, T., Kaetzel, C.S. and Davis, P.B. (1993) Gene-transfer into respiratory epithelial-cells by targeting the polymeric immunoglobulin receptor. *J. Clin. Invest.*, **92**, 2394–2400.
99. Hart, S.L., Harbottle, R.P., Cooper, R., Miller, A., Williamson, R. and Coutelle, C. (1995) Gene delivery and expression mediated by an integrin-binding peptide. *Gene Ther.*, **2**, 552–554.
100. Harbottle, R.P., Cooper, R.G., Hart, S.L., Ladhoff, A., McKay, T., Knight, A.M., Wagner, E., Miller, A.D. and Coutelle, C. (1998) An RGD-oligolysine peptide: a prototype construct for integrin-mediated gene delivery. *Hum. Gene Ther.*, **9**, 1037–1047.
101. Cooper, R.G., Harbottle, R.P., Schneider, H., Coutelle, C. and Miller, A.D. (1999) Peptide mini-vectors for gene delivery. *Angew. Chem. Intl. Ed.*, **38**, 1949–1952.
102. Wickham, T.J., Mathias, P., Cheresh, D.A. and Nemerow, G.R. (1993) Integrins $\alpha_V\beta_3$ and $\alpha_V\beta_5$ promote adenovirus internalization but not virus attachment. *Cell*, **73**, 309–319.
103. Bergelson, J.M., Shepley, M.P., Chan, B.M.C., Hemler, M.E. and Finberg, R.W. (1992) Identification of the integrin VLA-2 as a receptor for echovirus-1. *Science*, **255**, 1718–1720.
104. Logan, D., Abughazaleh, R., Blakemore, W., Curry, S., Jackson, T., King, A., Lea, S., Lewis, R., Newman, J., Parry, N., Rowlands, D., Stuart, D. and Fry, E. (1993) Structure of a major immunogenic site on foot-and-mouth disease virus. *Nature*, **362**, 566–568.
105. Hynes, R.O. (1992) Integrins: versatility, modulation, and signaling in cell adhesion. *Cell*, **69**, 11–25.
106. Pierschbacher, M.D. and Ruoslahti, E. (1984) Cell attachment activity of fibronectin can be duplicated by small synthetic fragments of the molecule. *Nature*, **309**, 30–33.
107. Haensler, J. and Szoka, F. C. Jr. (1993) Polyamidoamine cascade polymers mediate efficient transfection of cells in culture. *Bioconjugate Chem.*, **4**, 372–379.
108. Tang, M.X., Redemann, C.T. and Szoka, F.C. Jr. (1996) *In vitro* gene delivery by degraded polyamidoamine dendrimers. *Bioconjugate Chem.*, **7**, 703–714.
109. Boussif, O., Lezoualch, F., Zanta, M.A., Mergny, M.D., Scherman, D., Demeneix, B. and Behr, J.-P. (1995) A versatile vector for gene and oligonucleotide transfer into cells in culture and *in vivo*: polyethylenimine. *Proc. Natl Acad. Sci. USA*, **92**, 7297–7301.
110. Boussif, O., Zanta, M.A. and Behr, J.-P. (1996) Optimized galenics improve *in vitro* gene transfer with cationic molecules up to 1000-fold. *Gene Ther.*, **3**, 1074–1080.
111. Zanta, M.A., Boussif, O., Adib, A. and Behr, J.-P. (1997) *In vitro* gene delivery to hepatocytes with galactosylated polyethylenimine. *Bioconjugate Chem.*, **8**, 839–844.
112. Kircheis, R., Kichler, A., Wallner, G., Kursa, M., Ogris, M., Felzmann, T., Buchberger, M. and Wagner, E. (1997) Coupling of cell-binding ligands to polyethylenimine for targeted gene delivery. *Gene Ther.*, **4**, 409–418.
113. Erbacher, P., Remy, J.-S. and Behr, J.-P. (1999) Gene transfer with synthetic virus-like particles via the integrin-mediated endocytosis pathway. *Gene Ther.*, **6**, 138–145.

114. Abdallah, B., Hassan, A., Benoist, C., Goula, D., Behr, J.-P. and Demeneix, B.A. (1996) A powerful nonviral vector for *in vivo* gene transfer into the adult mammalian brain: polyethylenimine. *Hum. Gene Ther.*, **7**, 1947–1954.

115. Goula, D., Remy, J.-S., Erbacher, P., Wasowicz, M., Levi, G., Abdallah, B. and Demeneix, B.A. (1998) Size, diffusibility and transfection performance of linear PEI/DNA complexes in the mouse central nervous system. *Gene Ther.*, **5**, 712–717.

116. Ferrari, S., Moro, E., Pettenazzo, A., Behr, J.-P., Zacchello, F. and Scarpa, M. (1997) ExGen 500 is an efficient vector for gene delivery to lung epithelial cells *in vitro* and *in vivo*. *Gene Ther.*, **4**, 1100–1106.

117. Goula, D., Benoist, C., Mantero, S., Merlo, G., Levi, G. and Demeneix, B.A. (1998) Polyethylenimine-based intravenous delivery of transgenes to mouse lung. *Gene Ther.*, **5**, 1291–1295.

118. Kren, B.T., Bandyopadhyay, P. and Steer, C.J. (1998) *In vivo* site-directed mutagenesis of the factor IX gene by chimeric RNA/DNA oligonucleotides. *Nat. Med.*, **4**, 285–290.

119. Goldman, C.K., Soroceanu, L., Smith, N., Gillespie, G.Y., Shaw, W., Burgess, S., Bilbao, G. and Curiel, D.T. (1997) *In vitro* and *in vivo* gene delivery mediated by a synthetic polycationic amino polymer. *Nat. Biotech.*, **15**, 462–466.

120. Kabanov, A.V., Astafieva, I.V., Maksimova, I.V., Lukanidin, E.M., Georgiev, G.P. and Kabanov, V.A. (1993) Efficient transformation of mammalian-cells using DNA interpolyelectrolyte complexes with carbon-chain polycations. *Bioconjugate Chem.*, **4**, 448–454.

121. Astafieva, I., Maksimova, I., Lukanidin, E., Alakhov, V. and Kabanov, A. (1996) Enhancement of the polycation-mediated DNA uptake and cell transfection with pluronic P85 block copolymer. *FEBS Lett.*, **389**, 278–280.

122. Cherng, J.Y., van de Wetering, P., Talsma, H., Crommelin, D.J.A. and Hennink, W.E. (1996) Effect of size and serum proteins on transfection efficiency of poly((2-dimethylamino)ethyl methacrylate)-plasmid nanoparticles. *Pharmaceut. Res.*, **13**, 1038–1042.

123. van de Wetering, P., Cherng, J.Y., Talsma, H. and Hennink, W.E. (1997) Relation between transfection efficiency and cytotoxicity of poly((2-dimethylamino)ethyl methacrylate)/plasmid complexes. *J. Control. Rel.*, **49**, 59–69.

124. van de Wetering, P., Cherng, J.Y., Talsma, H., Crommelin, D.J.A. and Hennink, W.E. (1998) 2-(Dimethylamino)ethyl methacrylate based (co)polymers as gene transfer agents. *J. Control. Rel.*, **53**, 145–153.

125. Fritz, J.D., Herweijer, H., Zhang, G. and Wolff, J.A. (1996) Gene transfer into mammalian cells using histone-condensed plasmid DNA. *Hum. Gene Ther.*, **7**, 1395–1404.

126. Mistry, A.R., Falciola, L., Monaco, L., Tagliabue, R., Acerbis, G., Knight, A., Harbottle, R.P., Soria, M., Bianchi, M.E., Coutelle, C. and Hart, S.L. (1997) Recombinant HMG1 protein produced in *Pichia pastoris*: a nonviral gene delivery agent. *BioTechniques*, **22**, 718–729.

127. Gottschalk, S., Sparrow, J.T., Hauer, J., Mims, M.P., Leland, F.E., Woo, S.L.C. and Smith, L.C. (1996) A novel DNA-peptide complex for efficient gene transfer and expression in mammalian cells. *Gene Ther.*, **3**, 448–457.

128. Wadhwa, M.S., Collard, W.T., Adami, R.C., McKenzie, D.L. and Rice, K.G. (1997) Peptide-mediated gene delivery: Influence of peptide structure on gene expression. *Bioconjugate Chem.*, **8**, 81–88.

129. Legendre, J.Y., Trzeciak, A., Bohrmann, B., Deuschle, U., Kitas, E. and

Supersaxo, A. (1997) Dioleoylmelittin as a novel serum-insensitive reagent for efficient transfection of mammalian cells. *Bioconjugate Chem.*, **8**, 57–63.

130. Wyman, T.B., Nicol, F., Zelphati, O., Scaria, P.V., Plank, C. and Szoka, F.C. Jr. (1997) Design, synthesis and characterization of a cationic peptide that binds to nucleic acids and permeabilizes bilayers. *Biochemistry*, **36**, 3008–3017.

131. Murphy, J.E., Uno, T., Hamer, J.D., Cohen, F.E., Dwarki, V. and Zuckermann, R.N. (1998) A combinatorial approach to the discovery of efficient cationic peptoid reagents for gene delivery. *Proc. Natl Acad. Sci. USA*, **95**, 1517–1522.

132. Zuckermann, R.N., Kerr, J.M., Kent, S.B.H. and Moos, W.H. (1992) Efficient method for the preparation of peptoids [oligo(N-substituted glycines)] by submonomer solid-phase synthesis. *J. Am. Chem. Soc.*, **114**, 10646–10647.

133. Trubetskoy, V.S., Budker, V.G. Hanson, L.J., Slattum, P.M., Wolff, J.A. and Hagstrom, J.E. (1998) Self-assembly of DNA-polymer complexes using template polymerization. *Nucleic Acids Res.*, **26**, 4178–4185.

134. Uherek, C., Fominaya, J. and Wels, W. (1998) A modular DNA carrier protein based on the structure of diphtheria toxin mediates target cell-specific gene delivery. *J. Biol. Chem.*, **273**, 8835–8841.

135. Fominaya, J., Uherek, C. and Wels, W. (1998) A chimeric fusion protein containing transforming growth factor-α mediates gene transfer via binding to the EGF receptor. *Gene Ther.*, **5**, 521–530.

136. Sebestyén, M.G., Ludtke, J.J., Bassik, M.C., Zhang, G., Budker, V., Lukhtanov, E.A., Hagstrom, J.E. and Wolff, J.A. (1998) DNA vector chemistry: the covalent attachment of signal peptides to plasmid DNA. *Nat. Biotech.*, **16**, 80–85.

137. Zanta, M.A., Belguise-Valladier, P. and Behr, J.-P. (1999) Gene delivery: a single nuclear localization signal peptide is sufficient to carry DNA to the cell nucleus. *Proc. Natl Acad. Sci. USA*, **96**, 91–96.

138. Sanford, J.C., Klein, T.M., Wolf, E.D. and Allen, N. (1987) Delivery of substances into cells and tissues using a particle bombardment process. *Part. Sci. Technol.*, **5**, 27–37.

139. Williams, R.S., Johnston, S.A., Riedy, M., DeVit, M.J., McElligott, S.G. and Sanford, J.C. (1991) Introduction of foreign genes into tissues of living mice by DNA-coated microprojectiles. *Proc. Natl Acad. Sci. USA*, **88**, 2726–2730.

140. Shark, K.B., Smith, F.D., Harpending, P.R., Rasmussen, J.L, and Sanford, J.C. (1991) Biolistic transformation of a prokaryote, *Bacillus megaterium*. *Appl. Environ. Microbiol.*, **57**, 480–485.

141. Johnston, S.A., Anziano, P.Q., Shark, K., Sanford, J.C. and Butow, R.A. (1988) Mitochondrial transformation in yeast by bombardment with microprojectiles. *Science*, **240**, 1538–1541.

142. Daniell, H., Krishnan, M. and McFadden, B.F. (1991) Transient expression of beta-glucuronidase in different cellular compartments following Biolistic delivery of foreign DNA into wheat leaves and calli. *Plant Cell Rep.*, **9**, 615–619.

143. Mathiesen, I. (1999) Electropermeabilization of skeletal muscle enhances gene transfer *in vivo*. *Gene Ther.*, **6**, 508–514.

144. Rizzuto, G., Cappelletti, M., Maione, D. *et al.* (1999) Efficient and regulated erythropoietin production by naked DNA injection and muscle electroporation. *Proc. Natl Acad. Sci. USA*, **96**, 6417–6422.

145. Potter, H. (1988) Electroporation in biology: methods, applications and instrumentation. *Anal. Biochem.*, **174**, 361–373.

Chapter 5

Gene therapy for monogenic diseases

Georges Vassaux

5.1 Introduction

By definition, gene therapy involves transferring genes into cells, for
therapeutic purposes. In the context of monogenic diseases, gene therapy
approaches have invariably involved the delivery and expression of the
defective gene to the cells directly involved in the pathology or to their
precursors. Depending on the pathology considered, two different
strategies can be applied: *ex vivo* gene therapy or *in vivo* gene therapy.
The vector of choice to complement the defect has usually been the
retrovirus or other integrating vector, because of its capacity to integrate
its genetic information into the genome of the host cells. Thus, ideally, the
targeting and the transduction of the precursors of the cells affected by the
pathology should prevent the pathology. But behind this relatively simple
concept, many technical difficulties and challenges are hidden. They are
illustrated in this chapter by gene therapy for adenosine deaminase
deficiency (*ex vivo* gene therapy, *Figure 5.1*) and cystic fibrosis (*in vivo* gene
therapy, *Figure 5.2*).

5.2 Gene therapy for adenosine deaminase deficiency

5.2.1 Physiopathology

Adenosine deaminase (ADA) catalyses the irreversible deamination of
adenosine to inosine and 2′ deoxyinosine as part of the purine salvage
pathway. The enzyme is expressed in all body tissues and its action can be
seen as part of a detoxification process. Its deficiency results in the
syndrome of severe combined immunodeficiency (SCID). The condition
was first described in 1972 and ADA-deficient SCID became the first

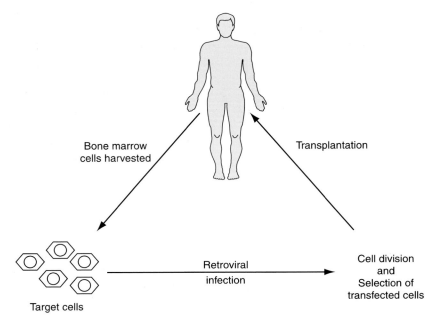

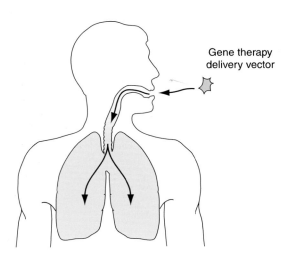

Figure 5.1. *Ex vivo* gene therapy for adenosine deaminase. The target cells (T lymphocytes or hematopoietic stem cells) are collected from the patient and infected *in vitro* with the therapeutic retrovirus containing the ADA cDNA. The transfected cells are then selected and transplanted back into the patient.

Figure 5.2. *In vivo* gene therapy strategy for cystic fibrosis. The gene therapy delivery vector containing the therapeutic gene (*CFTR* cDNA) is delivered directly to the lung by instillation or nebulization. As the target cells cannot be efficiently transduced by retroviruses, alternative nonintegrating vectors are used and the treatment has to be repeated in order to maintain therapeutic levels of *CFTR* expression.

primary immunodeficiency for which the molecular defect was identified. ADA–SCID is due to the accumulation of the products of degradation of adenosine, i.e. erythrocyte deoxyadenosine triphosphate (dATP), plasma and urinary deoxyadenosine (dAdo) and reduced levels of *S*-adenosyl-homocysteine hydrolase (SAHH) activity. These metabolites are primarily toxic to T lymphocytes and to a lesser extent to B lymphocytes, resulting in the absence of adequate numbers of functioning T and B cells.

The clinical manifestations of ADA–SCID include failure to thrive, diarrhea, recurrent pneumonia, skeletal abnormalities and opportunistic infections. The severity as well as the onset of the disease can be very variable, but generally the affected children die 1 to 2 years after birth from severe infections.

5.2.2 Therapeutic options

The treatment of choice for ADA–SCID is bone marrow transplantation (BMT). When a matched donor is available, the success rate of BMT can be as high as 90%. However, mortality and morbidity increase dramatically when transplantation is performed from a mismatched donor.

As an alternative treatment, extracellular enzyme replacement therapy was successfully demonstrated and an enzymatic preparation suitable for parenteral injection was developed: PEG–ADA, a bovine ADA preparation conjugated with polyethylene glycol (PEG). As dAdo and dATP are freely diffusible across the cellular membranes, the supply of an enzymatically active ADA preparation reduces the extracellular levels of dAdo and dATP. This process generates a concentration gradient that drives dAdo and dATP out of the cell, leading to a decrease in the intracellular concentration of the toxic metabolites of adenosine. In the majority of patients treated with PEG–ADA there has been an improvement in the biochemical abnormalities as well as in the immunological parameters, in particular in the number of functionally active lymphocytes. As a result, immunity is sufficient to protect against recurrent infections, repeated hospitalization and allow normal development of young patients. However, the treatment fails to stabilize the immune parameters in the long term. For ethical reasons, all children treated in the clinical trials of ADA–SCID gene therapy undertaken so far have also received PEG–ADA therapy.

5.2.3 Gene therapy for ADA–SCID

Since the ADA gene was cloned, ADA–SCID has been regarded as the perfect model disease for gene therapy. This is firstly due to the fact that the target tissue and cells are well defined. ADA is expressed in all body tissues but the pathology is largely due to damage to the immune system, particularly the T cells. These cells and their precursors in the hematopoietic lineage are easily accessible, can be genetically engineered to produce ADA *ex vivo* and reinfused into the patient. Moreover, based on results obtained from ADA-deficient patients treated by BMT, the genetically modified cells reintroduced to the patient are expected to have a selective advantage over the unmodified defective cells.

Another important consideration is the fact that the ADA complementary DNA (cDNA) to be delivered is relatively small (1.1 kb) and therefore can be accommodated in virtually all gene therapy delivery vectors and particularly retroviruses. The expression of the ADA gene does not need to be regulated by a cell-selective promoter and therefore, well characterized viral promoters driving high levels of constitutive expression can be used. In addition, studies performed on heterozygote carriers have shown that individuals with as little as 10% of normal ADA activity have no abnormality of immune function. Thus, on the basis of these observations, the treatment of a fraction of the cell population may be enough to provide improvement in the condition.

Pre clinical studies of gene therapy for ADA–SCID. The development of a new therapy follows a succession of steps, from *in vitro* experiments, to *in vivo* studies on animal models and, if successful, to clinical trials. Nearly all the studies on ADA–SCID gene therapy have used different versions of retroviral vectors as gene delivery systems. The characteristics of this type of vector are detailed in Chapter 3 but the main advantage of this system is that the transgene (in this case the ADA cDNA) will integrate into the genome of the recipient cell and therefore should be expressed as long as the transduced cell is alive. Two main cellular targets have been used: hematopoietic stem cells (HSCs) and T lymphocytes.

(i) Gene transfer to HSCs. The obvious target cells for gene therapy for ADA–SCID as well as for other conditions curable with BMT are the HSCs. Ideally, infection of these cells with a retrovirus containing the therapeutic transgene would result in the integration of this transgene in these progenitor cells, leading to the repopulation of all hematopoietic lineages with the genetically modified cells. Moreover, this approach would be a once-only procedure.

The initial studies involved transduction of murine marrow cells *ex vivo* under a variety of culture and infection conditions, followed by transplantation of these transduced cells into irradiated recipient mice. Many studies showed successful transfer as well as long-term *in vivo* expression of different transgenes including the ADA cDNA. For example, a particular study described the expression of the human ADA in all hematopoietic lineages of primary recipients 4 months after transplantation as well as expression of this transgene in the peripheral blood of secondary recipients. All together, these studies demonstrated the stability of transfer and expression of these murine progenitor cells. In an attempt to create an animal model of the disease, an ADA knockout mouse has also been generated but the phenotype of this mouse is very different from the human condition. In particular, no signs of immunodeficiency could be detected in these animals.

Following the success obtained with murine models, the next step was to perform the same type of experiments on larger animals (cats, dogs), including nonhuman primates (rhesus monkey). Coculture of early progenitors from the marrows of these animals with a retroviral packaging cell line and subsequent autologous transplantation of the retrovirally transduced cells resulted in a multi-lineage genetic modification that lasted more than 2 years after the gene transfer. However, in all these experiments the level of expression of the human ADA transgene was significantly lower than that achieved in the murine models. In some cases, the transgene stopped being expressed after 3 to 4 months.

As far as safety is concerned, no noticeable side-effects related to the gene transfer procedure were observed when retroviral vectors were derived from cell lines that were free of helper virus. However, lymphomas occurred in immunosuppressed monkeys exposed to murine amphotrophic murine helper virus, stressing the importance of using helper-free virus producing cell lines (see Chapter 3).

The final step in these preclinical studies of gene transfer to HSCs was to transduce human HSCs. This process required the isolation of human HSCs which were as primitive and totipotent as possible and the optimization of conditions to transduce these cells. Long-term bone marrow culture, an approach that had been successfully used for autologous BMT, was assayed. The results showed a high frequency of gene transfer by retroviruses. Another approach involved the isolation of cells from bone marrow on the basis of their expression of the CD34 molecule. CD34$^+$ cells consist of the earliest hematopoietic progenitors and are though to contain HSCs. *In vitro* studies demonstrated that these CD34$^+$ cells could be infected by retroviruses, although at a lower efficiency than that achieved with long-term bone marrow culture. An alternative source of HSCs was found in human umbilical cord blood. This blood is rich in progenitor cells and cord blood transfusion had already been used for a variety of hematopoietic diseases. Moreover, progenitors and CD34$^+$ cells from human umbilical cord blood can be transduced by retroviruses at frequencies similar to or greater than bone marrow-derived cells.

(ii) Gene transfer to T lymphocytes. The technical difficulties associated with the isolation, culture and transduction of HSCs led certain groups to target T lymphocytes for the correction of the ADA deficiency. Primary cultures of T lymphocytes from patients were retrovirally transduced and these ADA-expressing cells were shown to grow for a significantly longer period *in vitro*, compared to the untransduced cells. This observation confirmed the assumption that ADA transduced cells had a survival advantage. These results were extended *in vivo* by injection of

ADA-transduced peripheral blood lymphocytes from ADA–SCID patients into immunodeficient mice. Only the ADA-transduced cells were able to show long-term survival, accompanied by immunoglobulin production and development of antigen-specific T cells. Altogether, these results suggested that T-cell modification could reconstitute a certain degree of immune function.

Clinical trials of gene therapy for ADA deficiency

(i) T cell gene therapy. The first clinical trial of gene therapy for ADA was started on two girls in the USA in 1990. Both were on PEG–ADA therapy and had shown a good initial response to this treatment, followed by a deterioration of the lymphocyte number and response. The gene therapy protocol involved infection of peripheral T cells from the patient with a retrovirus containing human ADA cDNA in combination with recombinant human interleukine 2 (rIL-2) and an anti-CD3 antibody which both stimulated T-cell proliferation and thus improved retroviral transduction. The expanded T-cell population was then returned to the patients 9 to 12 days later. The procedure was repeated 12 times at regular intervals for each patient over a period of 18–24 months. From one round of transduction to the next, the efficiency of transduction varied from 0.1% to 10%. Polymerase chain reaction (PCR) analysis of peripheral cells performed up to 2 years after the last infusion showed that the retroviral sequences were present at a rate of 0.3 copy per cell in patient 1 while the level of transduced cells was only 0.1 to 1% in patient 2. Interestingly, these levels were stable over 2 years, a time period far longer than expected. The level of enzymatic ADA activity followed the results obtained by PCR, i.e. a significant rise in patient 1 but no change in patient 2. In both patients, the T-cell count rose rapidly after treatment and stabilized in the normal range for patient 1 and with a slight increase in patient 2. For both patients, cell-mediated immunity, T-cell immune response *in vitro* and humoral immune functions improved significantly. However, the continuous administration of PEG–ADA complicated the outcome of this trial. At present, the dose of PEG–ADA is being reduced.

(ii) T-cell depleted bone marrow and peripheral blood lymphocyte gene therapy. Another trial was performed on two patients with similar clinical conditions: the patients were treated with PEG–ADA and gene therapy was started when this treatment failed to have any effect on immunological parameters. The trial involved infusion of transduced T cells into patients, with increasing numbers of transduced HSCs later on. The two cell populations were transduced with slightly different retroviral vectors. This allowed an easy and precise evaluation of the efficiency of both approaches. Both patients received infusions of gene-modified peripheral blood lymphocytes and HSCs over a 2-year period. Initially, all vector

positive cells were derived from transduced peripheral blood lymphocytes. However, 1 year after the end of the treatment, the lymphocytes analyzed showed a bone marrow origin. Despite the presence of these transduced cells, the levels of ADA enzymatic activity remained low (less than 20% of the normal values). The immune reconstitution appeared to be more consistent than in the trial described in the previous paragraph with an increase in the absolute number of lymphocytes as well as in the number of active T cells, in both children.

(iii) CD34$^+$ cells gene therapy. Retroviral-mediated gene transfer to bone marrow CD34$^+$ cells was attemped on three ADA-deficient children in a once-only procedure. The gene transfer resulted in 5–12% of the cells being transfected *in vitro*. Transduced cells were detected 3 months after treatment (6 months after treatment for one patient), but no ADA gene expression was detected at any time.

In another trial, umbilical cord blood was the source of CD34$^+$ cells. Three infants were diagnosed prenatally and treated by autologous transplantation of retrovirally transduced CD34$^+$ cells. For all three patients, the PEG–ADA treatment started a few days after birth. Four years later, the number of gene-containing T lymphocytes has reached 1–10% whereas the frequency of other hematopoietic and lymphoid cells remained below 0.1%. However, cessation of the PEG–ADA treatment led to a decline in immune function despite the presence of ADA positive T lymphocytes.

5.2.4 Conclusion

Despite the initial optimism, no trial has yet achieved the objective of clinical cure. Very few patients have been treated and the long-term benefits of ADA–SCID on immune function remain to be clearly demonstrated. Moreover, the concomitant administration of PEG–ADA complicates the issue.

However, a number of encouraging points have been observed. It has been demonstrated that the transduction of HSCs is possible and that the T lymphocytes generated from these progenitors remain in the circulation for much longer periods than initially predicted. Importantly, no side-effect was reported as a result of gene transfer. The most significant concern was the contamination of the ADA-encoding retrovirus with wild type replication-competent retroviruses. But all the assays performed to detect wild type viruses were negative. Another safety aspect was the potential for insertional mutagenesis due to the random integration of the retrovirus-delivered transgene.

These initial studies have underlined the feasibility of gene therapy and have highlighted the problems that need to be overcome. Furthermore, they have given further insight into the biology of the hematopoietic cell lineage that may lead to improved second generation ADA–SCID gene therapy trial protocols.

5.3   Gene therapy for cystic fibrosis

5.3.1 Physiopathology and therapeutic options

Cystic fibrosis (CF) is a recessive disease affecting 1 in 2500 live births in Northern Europe and North America. Approximately 10% of the cases are diagnosed at birth, as the infant suffers intestinal blockage and the remainder within the next 5 years, as the affected child has increased susceptibility to respiratory infections.

The principal clinical problems of CF are lung damage and respiratory failure. They are the result of bacterial colonization of the thick mucus that accumulates in the lungs of the patients. These infections provoke cycles of inflammation that lead to brochiectasis and respiratory failure. The intestinal symptoms are generally milder and are of two types. The first results from perinatal obstruction of the small intestine and can usually be treated without surgery. The second is due to blockage and atrophy of the pancreatic ducts resulting in reduced secretion of pancreatic enzymes and reduced absorption of the gut content. Pancreatic insufficiency occurs in more than 80% of patients. Other features of the disease include liver problems due to the blockage of intrahepatic bile ducts (around 15% of patients) and azoospermia as a result of absence or obstruction of the vas deferens which renders CF males sterile.

In the 1930s the mean life expectancy of CF patients was 1 year and this has now been extended to 30 years in the UK. This increase is mainly due to the use of improved antibiotics in combination with intensive physiotherapy. In some rare cases heart and lung transplantations have been performed. In addition, the pancreatic insufficiency is controlled with pancreatic and dietary supplements.

In 1989, the gene responsible for CF was cloned and designated CFTR (cystic fibrosis transmembrane conductance regulator). The CFTR protein is a cAMP-mediated chloride channel which is nonfunctional in CF patients. The relationship between CFTR function and the pathology of CF is just beginning to be understood. One aspect is that the alteration in chloride transport affects the hydration of the cell surface, leading to reduced mucociliary clearance. Another aspect is that epithelial airways secrete a natural antibiotic, the action of which is sensitive to the intracellular salt content. This antibiotic activity is totally absent in CFTR⁻ cells, facilitating the colonization of the airways by opportunistic bacteria. Thus, the combination of reduced mucosal clearance with reduced defense against chest infections is currently thought to be responsible for the main CF pathology.

5.3.2 Gene therapy for CF

Considering the current state of the technology (and unlike gene therapy for ADA–SCID), expecting a cure from a gene therapy procedure for CF would be totally unrealistic. The main reasons are that retroviruses can not be used because they require cell proliferation for proviral integration and expression, and that the targets are post-mitotic, terminally differentiated lung epithelial cells. Moreover, the undifferentiated lung epithelial cell precursors have not yet been identified and characterized, despite 20 years of research. The alternative strategy of gene therapy for CF is to look for control of the condition rather than cure, by using alternative, non-integrating delivery vectors. This would involve repeated delivery of the CFTR gene to the target tissue for as long as the patient lives, in an attempt to stabilize the level of expression of the *CFTR* gene. Another difference between ADA–SCID and CF is in the nature of the disease which allows, in the case of ADA deficiency, an *ex vivo* approach rather than an *in vivo* delivery to CF patients. The case is complicated even more by the accumulation of thick mucus in the lungs of the CF patients that forms a barrier against any delivery vector and has been shown largely to inhibit transduction. On the other hand, a number of observations, obtained *in vivo* and *in vitro*, have suggested that levels of *CFTR* transgene expression as low as 5% should be sufficient to provide therapeutic benefits.

Many research groups have concentrated their efforts on the development of delivery systems and protocols suited to gene therapy for CF. The two main systems used are adenoviruses and liposomes. Another viral system, the adeno-associated virus (AAV), has also been considered but the limitation seems to be its packaging capacity that will just allow the packaging of the *CFTR* cDNA (4.5–5 kb).

Delivery vectors for gene therapy for CF

(i) Adenoviruses. These are ideal candidates for gene transfer to the airway. They have a natural tropism for the lungs and in particular, are capable of very efficiently infecting non-dividing cells. They can also be produced at very high titers. The wild type adenoviruses are associated with very minor pathologies. However, there is a possibility of both allergic and immune reactions, especially if the treatment is repeated.

In vitro studies examining adenoviral-mediated *CFTR* gene transfer were initially performed on CF cell lines, polarized epithelial monolayers derived from CF cell lines and freshly isolated human samples from CF patients, and proved to be successful. Subsequently, *in vivo* experiments were carried out on mice in which the *CFTR* gene had been knocked out (CF mice). One study failed to clearly demonstrate any correction of the chloride transport defect but other experiments carried out with improved adenoviral vectors showed up to 80% correction for a period of 3 weeks. Moreover, the human CFTR cDNA was successfully delivered to the lungs of cotton rats, with a sustained expression, after both single and repetitive administration. Adenovirus was also able to transfer the *CFTR* cDNA to primates. The expression was observed throughout the airways, with a patchy distribution.

(ii) Cationic liposomes. The generation of a liposome–DNA complex relies on the charge interaction between the negatively charged DNA and a mixture of lipids (cationic or neutral). The mechanism of cellular uptake is not yet clearly understood. Liposomes can be administered to the lungs as aerosols or by direct lavage. Intravenous injections can also be used but whether this leads to delivery to the surface cell epithelium remains to be demonstrated. Safety assessments have clearly demonstrated that liposome–DNA complexes are nontoxic, nonimmunogenic and do not deliver plasmid DNA to the gonads. However, unlike adenoviruses, they are relatively inefficient, mainly due to the fact that the DNA delivered by liposomes is degraded by lysosomes after the cellular uptake.

After many reports of correction of the chloride defect *in vitro*, using CF cell lines and primary cells from samples from CF patients, the feasibility of using liposomes to deliver *CFTR* cDNA *in vivo* was demonstrated using CF mice, originally in two studies. Liposome/CFTR expression plasmid complexes were delivered either by direct tracheal instillation or by aerosol. In both cases, cAMP-regulated Cl⁻ channel activity was introduced into the upper airways and one study managed to partially correct the Cl⁻ transport defect in the intestine, following rectal delivery. *In situ* hybridization was also used to demonstrate expression deep in the lungs. Delivery of the human *CFTR* plasmid DNA was also achieved in non-CF rodents. However, these observations were made using relatively large amounts of DNA and showed some variability in the correction.

CF clinical trials

(i) Nasal administration. The nasal cavity represents an ideal clinical test area to assess the safety and efficacy of *CFTR* gene transfer. The nasal epithelium demonstrates similar ion transport abnormalities to the lungs of CF patients and a functional test can be performed by direct measurement of the potential differences. With these noninvasive experiments, it is possible to reliably distinguish CF from non-CF patients and therefore to establish a scale of correction after the gene transfer procedure.

In the first clinical study published (1993), adenovirus-mediated *CFTR* cDNA transfer to the nose of three volunteer CF patients was reported. A degree of inflammation localized around the site of application was observed. *CFTR* transgene expression was detected in two patients and the chloride conductance was modified for 10 days after application. Another trial, involving adenovirus as the delivery vector, achieved an improvement in chloride secretion of 30% over a 2-week period without evidence of vector-induced epithelial damage. Finally, in a blind and placebo-controlled trial, 12 CF patients were recruited to receive one of four different doses of placebo or adenovirus to the nasal epithelium. Adenovirus-mediated *CFTR* mRNA was detected in five out of six patients who received the higher doses but no consistent changes in ion transport was observed. Mucosal inflammation was only observed at the highest dose (2×10^{10} viral plaque forming units [pfu]) and in two out of three patients.

A double-blind, placebo-controlled trial of liposome-mediated *CFTR* cDNA transfer was also performed and showed expression in five out of eight samples and a correction of the chloride conductance of 20% that had totally disappeared after 7 days. No safety problem was encountered.

(ii) Administration to the lower respiratory tract. As well as delivering *CFTR*-encoding adenovirus to the nose, one trial attempted to deliver the *CFTR* cDNA to the lungs by instillation through a bronchoscope. One bronchial sample was found to be positive for the normal *CFTR* protein but a patient who received the highest dose (10^{10} pfu) developed hypotension, fever and respiratory symptoms suggestive of an inflammatory reaction within the lungs. All clinical symptoms disappeared completely 1 month after the end of the treatment. A more recent trial, involving aerosol administration of a recombinant adenovirus expressing *CFTR* cDNA to the nasal epithelium as well as to the airways of CF patients demonstrated genetic transfer and *CFTR* expression without any acute toxic effects, but did not provide any data on ion transport improvement.

(iii) Conclusions from the current CF gene therapy trials. With the accumulation of data from many different preclinical and clinical trials, it is possible to draw a first conclusion on CF gene therapy. Irrespective of the delivery systems used, 50% of the samples show evidence of transgene *CFTR* expression in the nasal epithelium accompanied by around 30% of functional correction of the chloride conductance defect. In the lung, the only data available are on adenovirus-mediated delivery and show that the transgene mRNA has been detected in 20% of the samples and protein in 30%. With regard to safety, gene transfer appears to be safe in the nose and a correlation between the dose and the toxicity has been defined in the lungs for adenovirus. No data are available on liposome-mediated transfer to the lungs.

As far as protocols are concerned, the method of delivery of the vector to the lung seems to be of crucial importance and nebulization is the method that emerges from the different studies published. This technique has the potential disadvantage of mechanical stress applied to the liquid to nebulize and the risk of destruction of the delivery vector. Another complication lies in the presence of a very thick mucus in the lower respiratory tract that has been shown to inhibit transduction. To solve this problem, various mucolytic agents are currently under investigation.

Thus, it is reasonable to conclude that there is evidence for gene transfer as well as limited evidence for functional correction of the defect. At present, there are several CF gene therapy trials in progress, involving liposome-mediated delivery to the lungs or examining the effects of adenovirus administration to the nose and lungs. Finally, a trial using adeno-associated virus to deliver *CFTR* cDNA has been initiated.

5.4 Conclusion

These initial studies, involving ADA–SCID and CF have demonstrated the feasibility of gene therapy but have also been very useful in highlighting the problems that are still ahead. To tackle these problems, efficient *in vivo* delivery vectors will have to be developed to achieve levels of transduction comparable to those obtained with *ex vivo* gene therapy. New promoters, able to drive regulatable, cell-specific expression will also be needed. Finally, a more in-depth knowledge of the pathophysiology of the targeted disease is required to choose the right gene therapy strategic approach, in terms of delivery vector and protocol.

Further reading

Boucher, R.G. (1999) Status of gene therapy for cystic fibrosis lung disease. *J. Clin. Invest.*, **103**, 441–445.

Crystal, R.G. (1995) Transfer of genes to humans: early lessons and obstacles to success. *Science*, **270**, 404–410.

Hoogerbrugge *et al.* (1995) Gene therapy for adenosine deaminase deficiency. *Br. Med. Bull.*, **51**, 72–81.

Morsy, M.A. and Caskey, T. (1999) Expanded-capacity adenoviral vectors. The helper dependant vectors. *Mol. Med. Today*, **5**, 18–24.

Prince, H.M. (1998) Gene transfer: a review of methods and applications. *Pathology* **30**, 335–347.

Chapter 6

Gene therapy for multifactorial genetic disorders

Anne S. Rigg

6.1 Introduction

The most straightforward application of gene therapy is for the correction of a disease in which there is one easily identified genetic abnormality. This is unfortunately only the case for a few rare disorders such as adenosine deaminase deficiency. The major causes of mortality in the western world, namely coronary heart disease and cancer, are still poorly understood in terms of their genetic profiles. Even where the genetics have been studied the problem arises that these conditions are caused by a variety of genetic factors, often over a prolonged period of time and other environmental agents are also implicated. There are selected examples of diseases that have well-documented genetic mutations which illustrate the difficulties of promoting gene therapy for multifactorial conditions. Colorectal cancer, atherosclerotic vascular diseases and diabetes mellitus will be used to expand on this.

6.2 Colorectal cancer

6.2.1 Tumorigenesis

Colorectal cancer is the second most common malignancy in the western world and is one of the few tumors that has been extensively studied in terms of the process of tumorigenesis. Colorectal tumors are unusual in that they can be easily visualized and sampled via the colonoscope at all stages of development from small benign adenomas to invasive carcinomas. This has facilitated the investigation of the genotype of benign and malignant tumors. Vogelstein and his team at the Johns Hopkins University in Baltimore, USA proposed the now famous model of colorectal tumorigenesis. It was established early on that most, or all, of colorectal carcinomas arise from pre-existing adenomas. All colorectal tumors (benign and malignant) are monoclonal (i.e. arising from a single cell) as opposed to normal colonic epithelium which is polyclonal. Each time a tumor cell acquires a new genetic abnormality that confers a survival advantage there is a clonal expansion of this cell. It has been found that there is a median of 4–5 chromosomal losses per colorectal carcinoma cell. The progression from adenoma to carcinoma appears to occur by successive waves of clonal expansion as each new genetic event occurs. Tumors with mixtures of adenoma and carcinoma regions demonstrated that the carcinomatous cells had at least one more genetic abnormality than the adenomatous cells. It seems that it is the progressive accumulation of changes, rather than their order of occurrence with respect to each other that is important. Four to six independent steps are predicted to be necessary for tumorigenesis (*Figure 6.1*).

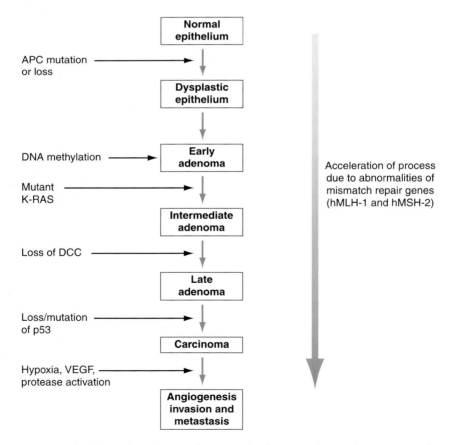

Figure 6.1. A scheme for the genetic events leading to colorectal tumorigenesis. Adapted from E.R. Fearon and B. Vogelstein (1990) *Cell*, **61,** 759–767. ©Cell Press.

6.2.2 Hereditary colorectal cancer syndromes

Knudson's 'two hit' hypothesis describes the fact that for a recessive genetic abnormality to affect the cellular phenotype the gene on both alleles would need to be altered/deleted. In the case of familial colorectal cancers affected individuals inherit a germline abnormality of one allele and then at a later point a genetic event affects the second allele. For sporadic tumors two independent genetic events need to occur to knock out the gene on both alleles. This may explain why patients with familial cancers often present at a younger age than those with sporadic cancers.

Familial adenomatous polyposis. Familial adenomatous polyposis (FAP) is an autosomal dominant condition which accounts for 0.5–1% of all colorectal cancers. Patients present with greater than 100 colorectal adenomas and these adenomatous polyps often undergo malignant change in the third to fifth decades of life. When extracolonic manifestations occur such as sebaceous cysts, congenital hypertrophy of retinal pigment epithelium and desmoid tumors this is described as Gardner's syndrome. It was noted that these colorectal adenomas had deletions of chromosome 5q. By restriction fragment length polymorphism and *in situ* hybridization the gene for FAP was localized to 5q21-5q22. In 1991 the tumor suppressor gene APC (adenomatous polyposis coli) was cloned from chromosome 5q21. It is the APC gene that is lost with the 5q deletion. The APC gene tends to be mutated on one allele and a deletion occurs on the other allele. It has been shown that APC gene inactivation is required for adenoma formation to occur. APC associates with certain cellular proteins, leading to their degradation. An example of this is β-catenin. When APC function is lost β-catenin accumulates and binds to T-cell factor-4 causing increased transcriptional activity of proto-oncogenes such as c-myc.

Hereditary non-polyposis colorectal cancer. Hereditary non-polyposis colorectal cancer (HNPCC) is another autosomal dominant condition characterized by colorectal carcinomas, predominantly on the right side of the colon. It can be divided into Lynch I syndrome (colorectal tumors only) and Lynch II syndrome (colorectal and extracolonic tumors). Most of these tumors have microsatellite instability and it is thought that this is due to mutations of mismatch repair genes. Two large studies of (HNPCC) families from New Zealand and Finland have demonstrated a linkage to microsatellite markers on chromosome 2p. A somatic mutation was identified that occurred in all HNPCC tumors, but only 15% of sporadic colorectal tumors. The mutations resulted in changes of length of the DNA tandem repeat sequence in microsatellite lesions known as replicative errors (RER) or microsatellite instability. Two mismatch repair genes have been identified, hMLH-1 and hMSH-2, of which germline mutations occur in HNPCC families. hMSH-2 mutations appear to confer a more severe phenotype with a higher rate of extracolonic tumors.

6.2.3 Sporadic colorectal cancers

Patients with nonhereditary colorectal cancers tend to present in the fifth and sixth decades. There has been much research into risk factors for colorectal cancer such as high levels of bile acids, low fibre diets, alkaline fecal pH, and low dietary calcium. Several types of genetic abnormalities occur in colorectal cells that can contribute to the process of tumorigenesis (*Table 6.1*).

Genetic alterations in oncogenes. Proto-oncogenes are a family of genes that normally promote cellular growth and differentiation. When altered, proto-oncogenes can become transforming oncogenes. This can occur through:

- point mutations that alter the function of the gene product to a transforming protein;
- gene amplification leading to increased production of the oncogene;
- abnormal regulation of the gene resulting in over-expression;
- gene rearrangements or translocations.

The most common example of activation of a proto-oncogene is the mutant *Kras* oncogene that is detected in 50% of carcinomas and 50% of adenomas greater than 1 cm in size. Interestingly, if the adenoma is less than 1cm only 10% harbour a *Kras* mutation. This would tend to suggest that a mutation of the *Kras* gene occurs in the cell of a pre-existing adenoma leading to clonal expansion, enlargement of the adenoma and dysplastic changes.

Oncogenic rearrangement has been described once for the *trk* oncogene in colorectal cancer. A few cases of specific gene amplification have been observed in colorectal cell lines and involve the *ERBB2, c-myc* or *c-myb* genes.

Table 6.1. Somatic mutations in colorectal cancer

Gene	Type of gene	Type of mutation
TP53	TSG	Point mutation, LOH
APC	TSG	Point mutation, LOH
DCC	TSG	Point mutation, insertion, Deletion, LOH
Kras	Proto-oncogene	Point mutation
Nras	Proto-oncogene	Point mutation
ERBB2	Proto-oncogene	Amplification, Hypomethylation
c-myc	Proto-oncogene	Amplification
myb	Proto-oncogene	Amplification

Tumor suppressor gene (TSG), Loss of heterozygosity (LOH)

Allelic losses and tumor suppressor genes. The second group of abnormalities detected involve the loss of regions of chromosomes. Analysis of restriction fragment length polymorphisms has enabled chromosomal loss to be examined on a smaller scale and to map precisely the loci that have lost heterozygosity (referred to as loss of heterozygosity analysis [LOH]). It is likely that the region lost contains a tumor suppressor gene if the chromosomal loss confers a more malignant phenotype on the cell. This has proved to be the case for the 75% of colorectal tumors that demonstrate a loss of part of the short arm of chromosome 17. It was discovered that 17p contains the tumor suppressor gene *TP53* whose protein product p53 is involved with arresting cell growth in response to DNA damage so that apoptosis can take place. Hence, the loss of p53 promotes tumor growth as abnormal cells no longer undergo apoptosis. Another example is the loss of 18q that occurs in 70% of carcinomas and 50% of late adenomas resulting in loss of the tumor suppressor gene DCC (deleted in colorectal cancer). DCC is thought to be a member of a family of cell adhesion molecules.

The difficulty with the inactivation of a tumor suppressor gene through loss of heterozygosity is that there is a second copy of the gene on the other allele. As described previously in the colorectal predisposition syndromes if there is a germline inactivation of one allele only one other event is required to produce tumorigenesis. However, for sporadic carcinomas two genetic events are needed to affect the tumor suppresor gene on both alleles. However, there are some exceptions to the rule. Essentially, clonal expansion of a particular cell should only happen if the genetic abnormality changes the phenotype of the cell in some way. It may be that this clonal expansion then provides plenty of cells in which a second event can take place. There are examples of a single allele mutation of a tumor suppressor gene producing a change in phenotype. Mutant p53 can act as a dominant negative and loss of DCC from one allele can cause a reduction in the density of the appropriate cell surface adhesion molecules and this in some way allows a decrease in growth-restraining signals.

Two other genes have been identified from regions that are commonly lost. The nm23 gene is located at 17q21 and may be related to the development of metastatic disease. The MCC gene lies close to the APC gene and is mutated in 15% of sporadic colorectal cancer, but is not associated with FAP. The function of MCC in tumorigenesis is as yet unknown.

DNA hypomethylation. Another possibility is that regions of cellular DNA become hypomethylated inhibiting chromosomal condensation and leading to mitotic non-disjunction and instability of the genome. Certainly even small adenomas have a third less methylation of their DNA than normal colorectal epithelial cells. If small regions of DNA are specifically hypomethylated inappropriate transcription of the genes at that locus can occur for example *ERBB2* and *myc* oncogenes.

Angiogenesis and metastasis in colorectal cancer. The mechanisms of metastasis are as yet not fully understood. Once a tumor reaches a critical size of 1 mm^3 it can no longer survive on oxygen and nutrients from the surrounding tissues. A central area of hypoxia develops and this seems to apply a selective pressure to the malignant cells such that those exhibiting reduced apoptotic potential (e.g. p53 mutation) tend to survive. One of the key events is the upregulation of the gene for vascular endothelial growth factor (VEGF) as its promoter region contains hypoxia induced factors. p53 and ras mutations also upregulate VEGF expression. Tumor endothelial cells proliferate rapidly and produce a new tumor vasculature. The vasculature is leaky and abnormal, allowing fibrinogen and plasminogen to leach into the tissues activating tissue factor and causing hypercoagulability. Plasminogen is also an activator of the extracellular matrix metalloproteinases which results in increased degradation of the extracellular matrix and promotes local invasion.

There are various groups of proteins that are implicated in the metastatic process. The MMPs degrade the extracellular matrix and activate other proteases. This allows tumor cells to migrate and intravasate into the blood vessels. Certain members of the MMP family are over-expressed in colorectal tumors. Molecules involved with cell–cell and cell–matrix interactions are crucial as they enable the tumor cell to migrate through tissues to a secondary site. Examples are the integrins, the cadherins and the immunoglobulin superfamily. In addition, tumor cells need to be able to survive in the circulation and indeed avoid immune detection at the tissue sites of disease. This is achieved through a variety of mechanisms in which components of the normal immunological pathways are down-regulated and tumor cells themselves have alterations that make them poorly immunogenic.

6.2.4 Cancer gene therapy

As has been demonstrated, colorectal cancer is a complex genetic disorder with a wide variety of abnormalites that probably develop over a long period of time. Therefore it would be naive to believe that the condition could be corrected with a single type of gene therapy. Which genetic abnormality should be targeted? This is a crucial and as yet unanswerable question as there are many potential targets. Often there is only a certain period of time during which a particular abnormality is driving clonal expansion before the next occurs. Thus the window of opportunity for correcting that specific abnormality may be limited.

Secondly, what is the best method of delivery of the gene therapy to the tumor cells? Again there is currently no accepted dogma. There are numerous vectors available but all have specific disadvantages to their use (as discussed in other chapters). The route of delivery and the vector used may ultimately be tailored to the site of disease, for example, retroviral vectors for brain lesions as retroviruses only infect dividing tissues, leaving the normal neural tissues unaffected. Below are detailed the main categories of gene therapy available for cancer gene therapy.

Tumor suppressor gene replacement. It has already been discussed that several tumor suppressor genes are lost or altered in cancers. It would therefore be sensible to attempt to replace the TSG function. To this end there have been many *in vitro*, *in vivo* and now a few clinical trials in which the aim is to replace the function of the TSG TP53. A trial of patients with lung cancers demonstrated that there was some suppression of tumor growth when a retrovirus expressing p53 was injected directly into tumors via the bronchoscope. The predominant problem with this approach is that it is impossible to genetically alter all of the tumor cells due to vector inefficiency. An alternative approach utilizing the abnormalities of p53 in tumor cells is the Onxy-015 virus. Normally, the adenoviral E1B protein binds to and inactivates p53 to block host apoptosis and thus allow efficient viral replication, ultimately resulting in cytolysis. In a novel approach an adenovirus lacking E1B was engineered. This E1B-attenuated adenovirus is unable to replicate in normal cells leaving them intact, but in tumor cells with abnormal p53 function viral replication is unchecked causing cytolysis. *In vivo* efficacy has been shown for human tumors xenografted into mice and clinical trials are in progress for a variety of tumor types including colorectal cancer.

Oncogenic inactivation by antisense technology. Antisense technology is the use of a nucleic acid that is complementary to a targeted region of DNA or RNA, and by hybridizing to its target prevents transcription/translation of that gene. This type of blockade of gene expression is suitable for genes that are over-expressed in tumor cells such as Kras and ERBB2. The antisense nucleic acid can be directed against DNA to form an irreversible triple helix, messenger RNA (mRNA) to block translation, transcription-associated proteins or can be a ribozyme with self-splicing activity that on binding to complementary mRNA causes cleavage of the molecule.

Antisense approaches have demonstrated an anti-proliferative effect for a variety of tumor types. However, there is discussion as to whether the oligonucleotides are acting as sequence-specific inhibitors or partly produce their anti-proliferative effect by binding to other molecules such as heparin. Additionally, oligonucleotides that contain CpG motifs are able to induce an immune response which itself may have an anti-tumor effect.

Genetic prodrug activation therapy (GPAT). GPAT targets a cytotoxic agent specifically to the tumor cells. Thus the often severe systemic toxicities associated with standard chemotherapy agents can be avoided. The strategy involves the transfer of a gene encoding a drug-metabolizing enzyme (suicide gene) to tumor cells. A nontoxic prodrug is then administered and is only converted to the cytotoxic agent in the cells producing the enzyme. The majority of enzymes utilized in this way are nonmammalian such as cytosine deaminase which converts 5-fluorocytosine to 5-fluorouracil and viral thymidine kinase which converts ganciclovir in conjunction with cellular enzymes to a toxic triphosphate form.

To allow selective gene transfer and expression in the tumor cells, and not normal cells, two approaches have been developed. The first is transcriptional targeting whereby tissue- or tumor-specific transcriptional regulatory elements (TREs: promoters and enhancers) are used to limit the expression of the suicide gene to tumor cells that express transcription factors that interact with those TREs. Examples of these TREs are the promoter regions from the tumor-associated genes carcinoembryonic antigen for colorectal cancer, alpha fetoprotein for hepatocellular carcinomas and ERBB2 for breast and colorectal cancers. Tissue-specific TREs include the tyrosinase promoter for melanomas.

The second targeting approach is to utilize a transductional characteristic of the target tissue. For example, retroviruses will only infect dividing cells and can therefore be used to deliver a suicide gene to brain tumors that are rapidly proliferating against a background of quiescent neural tissue. Clinical trials for patients with malignant brain tumors delivering thymidine kinase by a retrovirus and administering ganciclovir have demonstrated moderate success.

Genetic prodrug activation therapy will be discussed in more detail in Chapter 8.

Immunotherapy. Cancers are poorly immunogenic. One of the reasons for this is that tumors fail to produce a T-helper cell response. These T-helper cells would normally release cytokines stimulating cytolytic T-cells to destroy the tumor cells. A way of bypassing this process is to provide the cytokine directly. Unfortunately, the systemic administration of high doses of interleukin-2 had only limited efficacy for a few cancers (renal cell carcinoma and melanoma) at the expense of high systemic toxicity. The challenge then was to produce small doses of cytokines at the site of the tumor itself. This was attempted by adoptive transfer of cytokine genes (TNF and interleukin-2) to tumor-infiltrating lymphocytes, tumor-associated lymphocytes and lymphokine-activated killer cells *ex vivo* and then reinfusing them into the patient. Alternatively, tumor cells could be transduced with cytokine genes *ex vivo*, the tumor cells irradiated to eliminate malignant activity and reintroduced into the host.

These approaches lack tumor specificity and do not solve the defects of antigen processing, presentation and lack of co-stimulatory signals for T-cell activation. Thus attempts have been made to introduce genes encoding allogeneic HLA and co-stimulatory molecules into tumor cells to elicit a response. There are also efforts to propagate effective T-cell responses to oncoproteins such as ras and c-erb-b2. T-cells from healthy individuals and cancer patients can distinguish between wildtype and mutant p21ras which differ by only a single point mutation. Synthetic mutant ras peptides containing putative T-cell epitopes have been pulsed onto professional antigen presenting cells and injected into patients with pancreatic cancer harbouring the same ras mutation. Two of five patients treated showed evidence of T-cell responses. Dendritic cells are antigen presenting cells and can now be expanded *ex vivo* from peripheral blood or bone marrow. These dendritic cells are able to take up exogenous antigen such as tumor proteins or can be transduced with genes encoding tumor antigens. The modified dendritic cells can then be returned to the patient to induce an efficient immune response. Of 16 patients with melanoma treated in this way five had an objective tumor response and none had side-effects.

Polynucleotide vaccinations can deliver genes that express unique oncoproteins such as p53 and kras into cells at the site of vaccination which might then elicit an efficient MHC class I CD8+ response rather than the less effective class II CD4+ response. DNA vaccination has also been tried and this has the advantages of ease of production and purity. A murine study showed that intramuscular injection of a synthetic plasmid expressing the human CEA cDNA could stimulate an antigen-specific humoral response and provide protection against tumor challenge with CEA-expressing colorectal carcinoma cells.

Multidrug resistance genes. A major disadvantage of current cytotoxic agents for cancer is the development of tumor resistance to the drugs. The multidrug resistance gene 1 (MDR1) produces P-glycoprotein and this acts as a cellular efflux pump accounting for tumor cell resistance to certain chemotherapy agents. Transfer of the MDR1 gene to normal bone marrow stem cells *ex vivo* means that on resettlement in the bone marrow these cells will be chemotherapy-resistant and higher doses of the drugs can be given to kill the tumor avoiding the problems of profound myelosuppression. Similarly, marrow cells transduced to express dihydrofolate reductase are protected against methotrexate toxicity. Clinical trials are now in progress with MDR1 for breast and ovarian cancer patients receiving paclitaxel. There are concerns such as the level of nonhematological toxicity with the higher drug doses and the possiblity of transducing malignant cells in the bone marrow with the resistance gene.

Vascular and stromal targeting. Many groups have investigated possible strategies to reduce tumor angiogenesis and invasiveness. In principle there are three main strategies:

- suppression of positive regulators;
- induction of negative regulators;
- suppression of receptor expression on endothelial cells.

Suppression of positive regulators such as VEGF can be achieved with monoclonal antibodies or antisense molecules. Levels of the regulator acidic fibroblast growth factor were reduced by a specific antisense ribozyme to fibroblast growth factor binding protein. Several natural anti-angiogenic agents exist, for example angiostatin, endostatin, thrombo-spondin-1, platelet factor 4 (PF4) and the tissue inhibitors of matrix metalloproteinases (TIMPs). Genes encoding these agents can be transduced into tumor cells. PF4 has demonstrated inhibition of proliferation of endothelial cells adjacent to glioma cells transduced with the PF4 gene *in vitro* and *in vivo*. TIMP2 has also inhibited blood vessel formation and invasion in transduced melanoma cell lines, and angiostatin in fibrosarcomas.

Experiments studying the effect of blockade of angiogenesis-promoting receptors show a reduction in angiogenesis, for example antagonists to the VEGF receptors, integrin $\alpha v \beta 3$ receptors and VE-cadherin.

6.3 Atherosclerotic vascular diseases

Blood vessels are composed of a layer of endothelial cells lining the lumen. Beneath the endothelial cells are vascular smooth muscle cells and beyond these a connective tissue adventitia through which run the vasa vasorum. The endothelial cells provide an interface between the blood and the vessel wall, while the smooth muscle layer controls contractile tone of the vessel in response to vasoactive substances.

6.3.1 Atherosclerosis

Atherosclerosis is a complex disorder resulting from an interplay of multiple inherited and environmental factors. In the western world it is the major cause of morbidity and mortality from coronary heart disease, cerebrovascular accidents ('strokes') and peripheral vascular disease. Atherosclerosis can be defined as a localized inflammatory fibroproliferative response to endothelial injury. The injury may be mechanical, viral, toxin-induced or immunological in origin. Once the endothelial cells have been damaged macrophages/monocytes adhere to the endothelial surface and invade the subendothelial space. These macrophages/monocytes ingest lipoproteins and become lipid-filled cells bulging under the endothelial surface and producing the appearance of a fatty streak. The lipoproteins ingested are in the oxidized form which seems to switch on their atherogenic potential. There appears to be no regulation to the macrophage/monocyte ingestion of these oxidised lipoproteins so the process continues until the cell is overladen. Interestingly, most of the environmental risk factors linked to atherosclerosis are thought to produce their effect by providing an oxidative stress (*Table 6.2*). Smooth muscle cells begin to proliferate in the vicinity and platelet thrombi accumulate on the endothelial surface. The so-called 'plaque' has now been formed. Platelet accumulation is enhanced if the plaque ruptures. The plaque becomes further developed by the recruitment of other cells to the site and the laying down of fibrous tissue.

Plaques can exert a pathological effect by direct occlusion of the vessel (e.g. myocardial ischemia, leg muscle ischemia), by releasing fragments of thrombi into the circulation known as emboli or by weakening the vessel wall resulting in aneurysm formation and rupture. Of note, the process of plaque formation can be reversible and this has been the basis of treatment.

6.3.2 Genetic mechanisms for the development of atherosclerosis

The genetic basis for the development of atheromatous plaques is highly complex. There are several factors that have been shown to promote intimal thickening. One of these is platelet-derived growth factor which also causes smooth muscle cell proliferation and synthesis of the extracellular matrix. TGF-β is involved in matrix secretion and angiotensin II promotes smooth muscle cell survial and proliferation. Down-regulation of enzymes such as endothelial nitric oxide synthase (eNOS) allows intimal thickening, inhibits platelet aggregation and causes vasoconstriction. Proto-oncogenes and tumor suppressor genes that normally regulate the cell cycle are also implicated. Upregulation of the former and down-regulation of the latter encourages cells to pass through the G1/S checkpoint of the cell cycle and begin proliferating. There is also evidence of cytokine involvement in the process for example interleukin-1, TNF-α and γIFN. Hyperlipidemic conditions predispose to atherosclerosis which can be due to a variety of genetic lesions.

6.3.3 Treatment for atherosclerotic diseases

Standard treatments involve the correction of risk factors such as weight loss, cessation of smoking, antihypertensive agents and lipid-lowering

Table 6.2. Risk factors for coronary heart disease (CHD)

Risk factor	
Tobacco smoking	Modifiable
Hyperlipidemia	Modifiable
Hypertension	Modifiable
Obesity	Modifiable
Lack of physical activity	Modifiable
Family history of CHD	Nonmodifiable
Diabetes mellitus	Nonmodifiable
Males or post-menopausal women	Nonmodifiable
Increasing age	Nonmodifiable

agents. For the acute cardiac event (heart attack) treatment with anticoagulants, and anti-platelet drugs and plasminogen activating drugs is indicated. Surgical intervention takes the form of bypass grafting and balloon angioplasty of stenotic lesions to provide symptomatic relief. Of note, a common event after angioplasty is restenosis of the vessel. This is thought to occur due to intimal hyperplasia, adventitial thickening and healing of fissures that are produced in the plaques on angioplasty.

6.3.4 Gene therapy for atherosclerosis

Gene therapy has commenced for the treatment of atherosclerotic vessels. Studies using marker genes have shown that there can be successful gene transfer to the endothelial and smooth muscle cells via intraluminal delivery. There have been many different approaches which have been broadly outlined in *Table 6.3*.

6.4 Diabetes mellitus

Diabetes mellitus is a disease characterized by hyperglycemia. There are two distinct types. Insulin dependent diabetes mellitus (IDDM) is an autoimmune disorder that typically affects younger people (0.3% of the UK population). These patients appear to have an inherited predisposition and the overt disease is then triggered by an environmental event such as a viral infection. The pancreatic β-cells, that are the source of insulin, are destroyed over a period of time resulting in severe insulin deficiency. The treatment therefore is insulin replacement by injection. The second type is noninsulin dependent diabetes mellitus (NIDDM) which tends to be a disorder of middle age (3% of the UK population). Again, there is thought to be an inherited component. Environmental factors such as obesity and lack of exercise are risk indicators. In this case the disease is characterized by insulin resistance as well as insulin deficiency. Standard treatments include diet, oral antihyperglyemic agents and insulin injections for the most severe cases. IDDM and NIDDM can cause long-term damage to other organs of the body such as nephropathy, retinopathy, neuropathy and advanced atherosclerosis of the blood vessels. These complications have been shown to be more common if a patient's diabetes is poorly controlled. However, it is often extremely difficult to achieve strict glycemic control with pharmacological agents. So although there are reliable treatments available for diabetes mellitus there is an argument for considering new genetic therapies that might provide better physiological control of blood glucose levels.

6.4.1 The genetics of insulin production

The human insulin gene is located on chromosome 11p. It is expressed only in the β-cells of the pancreatic islets of Langerhans. The promoter

Table 6.3. Gene therapy strategies for atherosclerotic vascular disorders

Strategy title	Gene/treatment	Mechanism	*In vivo*/clinical data
Genetic prodrug activation therapy	HSVTK gene and ganciclovir	Endothelial and smooth muscle cells transduced with suicide gene. Prodrug administered and only transduced cells killed	HSVTK delivered to balloon-injured rat and rabbit arteries. Resulted in reduced smooth muscle cell proliferation on administration of ganciclovir.
Promotion of collateral blood vessel development	VEGF Fibroblast growth factor	Both are potent pro-angiogenic factors	VEGF plasmid i.m. for patients with peripheral vascular disease led to symptomatic and angiographic improvements. hFGF plasmid into rat left ventricular wall promoted capillary density in the myocardium.
Cytostatic (reduced proliferation but maintenance of cell viability)	Endothelial nitric oxide synthase (eNOS)	Decreases proliferation, inhibits platelet aggregation and vasodilates	eNOS transferred to the endothelium after balloon injury, reduced atherosclerosis and limited restenosis.
Promotion of apoptosis	(1) Insertion of tumor suppressor genes (p53, Rb, p21) (2) Inactivation of proto-oncogenes by dominant negative mutants or antisense (*c-myc, c-myb, ras*)	Reduced entry of cells into cell cycle and promotion of apoptosis	Use of a nonphosphorylatable form of Rb blocks restenosis *in vivo*. Antisense to *c-myc* embedded in slow-release polymer and coated around the artery inhibited smooth muscle cell migration.
Familial hypercholesterolemia	Low density lipoprotein receptor (LDLR)	Disease caused by deficiency of LDLR therefore replacement strategies considered	*Ex vivo* transduction of hyperlipidemic rabbit hepatocytes with a retrovirus encoding LDLR gene. Hepatocytes then reinfused into the liver via the portal vein producing a lowering of serum cholesterol lasting more than 120 days. Now in clinical trial.
Inhibition of cell-cell and cell-matrix interactions	Anti-integrins Tissue inhibitors of matrix metalloproteinases Soluble VCAM-1	Disruption of cell adhesion and migration abilities	Adenoviral vector encoding soluble VCAM-1 acts as a competitive inhibitor for the binding of cellular VCAM-1 to VLA4 on the surface of monocytes thus preventing them from invading the endothelium.
Plasminogen activators for arterial thrombosis	Urokinase-type plasminogen activator Tissue plasminogen activator		

HSVTK, herpes simplex virus thymidine kinase; i.m., intramuscular; VEGF, vascular endothelial growth factor; Rb, retinoblastoma; VCAM-1, vascular cell adhesion molecular-1; VLA4, very late antigen 4 region has been characterized and contains many transcription factor binding sites. Several of these sites bind transcription factors that are exclusive to β-cells. Upstream of the promoter is a hormone response element and close to this is the locus for IDDM2. Once transcribed and translated preproinsulin is folded within the endoplasmic reticulum and then packaged into secretory granules of proinsulin by the transgolgi network. An increase in intracellular calcium is the stimulus for release of the granules. Proinsulin is then cleaved by two endopeptidases. β-cells take up glucose so that the intracellular concentration is equivalent to that in the systemic circulation. This is performed by a glucose transporter GLUT2. The intracellular glucose is then phosphorylated by glucokinase and the more glucose present the higher the metabolic rate leading to an increase in the ATP/ADP ratio. This results in closure of the ATP-dependent K^+ dependent pump. The subsequent depolarization activates a voltage-dependent Ca^+ channel and thus the intracellular calcium level rises stimulating proinsulin release.

6.4.2 Ex vivo *gene therapy approaches for diabetes mellitus*

The process of glucose homeostasis is a complex interaction between insulin, other peptide hormones and the sympathetic nervous system. This makes it very difficult to attempt to mimic. Most attention has been focused on genetically modifying cells *ex vivo* and then implanting them into the body to produce insulin. These cells could range from β-cells with a physiological control mechanism to sense levels of blood glucose and respond accordingly to non-β-cells cell type that could provide a constant background level of insulin secretion.

Immortalized β-cell lines lose their glucose responsiveness with time. If then transfected to express GLUT2 there was an appropriate insulin secretion in response to changing glucose concentrations. Researchers then began to concentrate on a system whereby β-cell secretion of insulin could be controlled by a genetic switch that could be turned on and off as required. β-cells can be immortalized by the introduction of the SV40 large T-antigen (Tag – a viral oncogene). Tag acts by sequestering p53 and retinoblastoma proteins that normally control entry into the cell cycle and apoptosis. Tag was joined to the rat insulin gene promoter (RIP) and injected into mouse embryos to generate offspring with the Tag expressed only in the β-cells by virtue of the RIP driving it. However, β-cell tumors eventually developed due to the oncogenic properties of Tag. The system was further modified by placing the SV40 Tag under the control of the bacterial tetracyclin operon regulatory system (tet). A series of experiments were conducted to produce a double transgenic mouse that had tetracyclin-controlled activation of Tag exclusively in the β-cells of the pancreas. The mice were allowed to develop β-cells tumors due to the oncogenic properties of Tag. The tumors were then harvested and grown in culture. These cells were then implanted into a diabetic mouse model and resulted in normalization of blood glucose levels within two weeks. If no tetracyclin was given the implanted cells continued to proliferate under the influence of Tag and the mouse died of profound hypoglycemia. However, if a slow-release pellet of tetracyclin was administered the mice had normal blood glucose levels for up to 4 months.

There is some knowledge of the process by which islet cells develop embryologically, but little is known about how islet cells are replaced in the adult. There must be islet stem cells present but it is currently impossible to identify them. Theoretically, if they could be identified they could be transfected with the appropriate factors to make them differentiate into β-cells. An alternative is to consider the genetic modification of non-β-cells with the insulin gene. This could be done for an individual diabetic patient by removing tissue from an accessible site such as skin, transducing it *in vitro* with the insulin gene and reimplanting into the patient (syngeneic). Allogeneic procedures could be done in an identical manner, but then involve the problem of rejection when transplanted into the patient. One problem is that non-neuroendocrine cells lack the secretory pathway and the endopeptidases needed to activate insulin. One way found to circumvent this difficulty was to mutate the proinsulin cleavage site so that it could be cleaved by the enzyme furin that is present in most cells. It was demonstrated that for a variety of cell lines transfected with this mutant insulin cDNA the level of insulin produced was directly correlated to the furin level of the cells. Of note, myoblasts transfected with the mutant insulin cDNA could secrete fully processed insulin and interest has now been focused on using transduced myoblast implantations to provide longterm background secretion of insulin in patients. It may eventually be possible to co-transfect with the appropriate transcription factors and enzymes to produce a homeostatic system.

6.4.3 In vivo *gene therapy for diabetes mellitus*

As yet the exact inherited genetic anormalites for IDDM and NIDDM are not known so gene therapy has really been focused on mechanisms of augmenting insulin production by engineering cells to produce insulin *ex vivo* or *in vivo*. Rodent experiments have shown that the rat insulin gene with the appropriate regulatory sequences can be delivered to the liver and be taken up by hepatocytes and expressed to produce insulin secretion. One method using a retroviral vector encoding the rat insulin gene injected into the portal vein showed that when the rats were subsequently rendered diabetic the ectopic expression of insulin by the liver prevented hyperglycemia and death.

6.5 Summary

It has been demonstrated that the commonest diseases of the western world have a multigenic pathogenesis. This makes a single gene correction inappropriate. New genetic abnormalities associated with the three examples discussed (colorectal cancer, atherosclerotic vascular disease and diabetes mellitus) are still being discovered. There are many challenges ahead to make gene therapy a viable option for these disorders. The most useful genetic targets and the optimal time for the administration of gene therapy during each disease process need to be identified. Despite this there is considerable optimism for the future of gene therapy in this setting.

Further reading

Docherty, K. (1997) Gene therapy for diabetes mellitus. *Clin. Sci.*, **92**, 321–330.

Fearon, R.R. and Vogelstein, B. (1990) A genetic model for colorectal tumorigenesis. *Cell*, **61**, 759–767.

French, B.A. (1996) Cardiovascular disease In: *Gene Therapy* (eds N.R. Lemoine and D.H. Cooper) BIOS Scientific Publishers Ltd, Oxford.

Kinzler, K.W. and Vogelstein, B. (1996) Lessons from hereditary colorectal cancer. *Cell*, **87**, 159–170.

Levine, F. and Leibowitz, G. (1999) Towards gene therapy of diabetes mellitus. *Mol. Med. Today*, **5**, 165–171.

Tomlinson, S., Heagerty, A.M. and Weetman, A.P. (eds) (1997) *Mechanisms of Disease. An Introduction to Clinical Science*. Cambridge University Press, Cambridge.

Chapter 7

Gene therapy for infectious diseases

Sarah Fidler and Jonathan Weber

7.1 Introduction

Infection of human cells with any pathogen that inserts its own genetic material into the host cell genome can be considered to be an acquired genetic defect, and therefore potentially treatable through gene therapy. This can either directly target the foreign gene, or manipulate the immune response via genetic means.

Conventional therapy for infectious diseases has historically employed two approaches; first; by disease prevention, via vaccination and public health, such as in the prevention of polio virus infection, or second; by the treatment of the established infection with antimicrobial chemotherapy.

The need for alternative therapies to treat certain infectious diseases has stemmed from the combination of a lack of sufficient efficacious agents or vaccines, in addition to the emergence of widespread resistance to antimicrobial chemotherapy. Infections that are potentially susceptible to gene therapy are those causing chronic disease and premature death, where no current adequate treatment or vaccination exists: these include; herpes simplex virus (HSV), cytomegalovirus (CMV), hepatitis B virus (HBV), hepatitis C virus (HCV), human T-cell leukemia virus (HTLV-1), tuberculosis (TB) and human immunodeficiency virus (HIV-1). All are highly pathogenic, potentially fatal, and infect vast numbers of people (the estimated number of HIV infected individuals world-wide is approaching 40 million, and tuberculosis approximately 60 million).

HIV-1 is a growing pandemic, and is an example of a highly variable organism, with the ability to become resistant to current therapies even when used in combination, and with a long clinically latent period which facilitates efficient transmission. For these reasons the majority of international research on gene therapy in the arena of infectious diseases has been focused on HIV.

Gene therapy could affect the infectious process either by preventing the new infection of uninfected target cells, or by blocking the production of newly synthesized virus from infected cells. For any infection the natural host immune responses attempt to suppress the infecting organism. Augmentation of this naturally occurring process by various genetic manipulations of the immune response may influence the outcome of infection. Alternatively, genetic means have been developed that attack the invading pathogen directly, and thereby prevent the establishment or dissemination of infection. The use of gene therapy can therefore be divided into antiviral or immunomodulatory, and these two aspects will be treated separately.

7.2 Strategies for antiviral gene therapy

Three principal strategies govern the development of gene therapy of infectious diseases, as illustrated in *Figure 7.1*:

(1) protect uninfected cells from viral entry or insertion of viral DNA into the host cell genome;

(2) a direct antiviral effect active against virally infected cells only, sparing uninfected cells;

(3) specifically killing virally infected cells.

7.2.1 The application of gene therapy to infection with HIV

Understanding the viral life cycle enables potential sites for the application of gene therapy to be considered. *Figure 7.2* illustrates a schematic diagram of the HIV life cycle. Like all viruses interaction with target cell surface membrane receptors precedes cell entry. Following entry viral RNA is reverse transcribed into DNA and integrated into host cell genome. Here it resides until cellular activation, where upon viral transcription is initiated and new virions are synthesized. Gene therapy can potentially influence any of the stages described, but can best be divided into pre- and post-integration steps. Prevention of integration of viral genome into the host cell will prevent infection of that cell and thereby interrupt pathogenesis, whilst prevention of post-integrational events interrupts the formation of new viable infectious virions. Prior to discussing antiviral gene therapy for HIV infection, the normal replication life cycle of HIV-1 will first be considered.

7.2.2 HIV life cycle

HIV-1 is a retrovirus, a member of the complex group of Lentiviruses, probably originating from a primate species. It contains a double strand of viral RNA surrounded by matrix and capsid, enveloped in a complex glycoprotein comprising of two parts; gp120 and gp41.

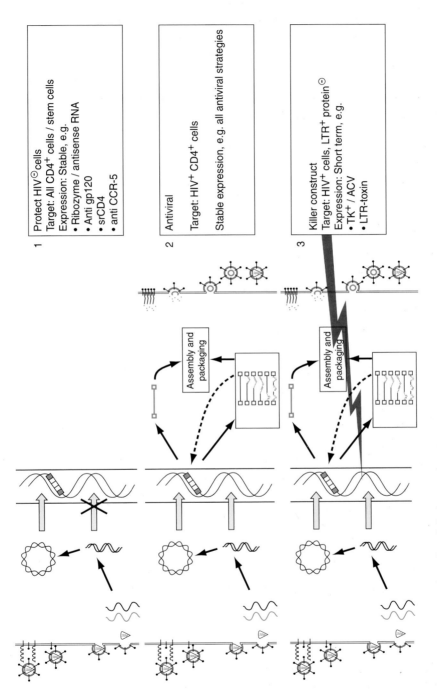

1   Protect HIV⊖ cells
    Target: All CD4+ cells / stem cells
    Expression: Stable, e.g.
    • Ribozyme / antisense RNA
    • Anti gp120
    • srCD4
    • anti CCR-5

2   Antiviral
    Target: HIV+ CD4+ cells
    Stable expression, e.g. all antiviral strategies

3   Killer construct
    Target: HIV+ cells, LTR+ protein ⊕
    Expression: Short term, e.g.
    • TK+ / ACV
    • LTR-toxin

Figure 7.1. Strategies of antiviral gene therapy.

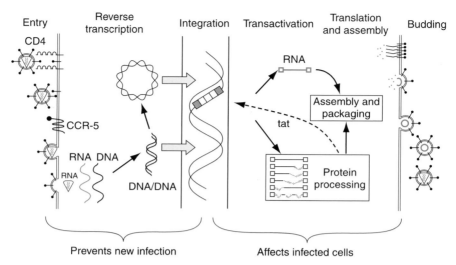

Figure 7.2. HIV life cycle.

HIV-1 has a complex viral genome, containing within its 9 kilobase RNA sequence three main genes; *gag*, *env*, and *pol*, encoding the core nucleocapsid protein; surface envelope glycoprotein gp120, and the viral enzymes such as reverse transcriptase, respectively. In addition there are six other genes that have a regulatory function and control replication and infectivity. *Figure 7.3* illustrates a diagrammatic representation of an HIV genome and summarizes the principal functions of each viral gene.

The human CD4 molecule represents the cell receptor for HIV-1, where part of the viral envelope glycoprotein gp120, the V3 loop, a highly variable region of the HIV envelope, interacts with the CD2 domains of CD4. CD4 is a cell surface receptor expressed on the surface of cells of the immune system, including CD4 + T cells, monocytes, macrophages, dendritic cells, and microglial cells of the brain. Its natural ligand is part of the self MHC class II molecule, whose function is to co-stimulate T-cell activation upon interaction with a specific peptide loaded MHC class II molecule with the T-cell receptor. A schematic representation of the HIV life cycle is illustrated in *Figure 7.2*. To gain entry into a susceptible CD4 bearing target cell, HIV requires in addition to CD4 the interaction with a second co-receptor, a member of the chemokine receptor superfamily, CCR-5 or CXCR4.

The importance of this second family of receptors in allowing viral entry was recently substantiated following the observation that a single gene deletion (32 base pairs), of part of the gene encoding the chemokine receptor conferred protection against *in vitro* challenge with HIV-1, and afforded protection clinically from established infection following sexual exposure to HIV-1 [1]. After uncoating, reverse transcription of viral RNA commences resulting in the production of double stranded DNA. In the presence of appropriate host cell factors, double stranded viral DNA

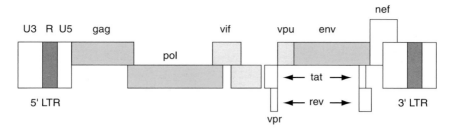

Figure 7.3. HIV-1 genomic structure. The 5′ and 3′ long terminal repeats (LTR) contain binding sites for host transcription factors. The function of the nine known genes of HIV-1 are summarized.

Gene	Function
gag	Encodes for nucleocapsid core proteins
pol	Encodes for reverse transcriptase, integrase, protease and RNase
vif	Promotes viral infectivity
vpr	Nuclear targeting
vpu	Required for virion budding
env	Encodes for the surface-coat proteins
tat	Amplifies viral gene expression
rev	Regulates structural gene expression
nef	Enhances viral infectivity

is inserted into the host cell genome where it remains dormant as the HIV-1 provirus. After entry into a CD4 expressing target cell HIV may establish a persistent or latent form of infection, the basis of which maybe determined by the state of activation of the target cell. Upon host cell activation the expression of the HIV-1 gene is stimulated initially by the action of host transcription factors with binding sites in the long terminal repeat (LTR), which then leads to sequential viral mRNAs. Viral proteins are then produced and assembled into new virions. Free HIV-1 virions produced by budding from the host cell are then free to reinitiate the retroviral life cycle and infect new CD4 expressing target cells.

Recent studies have identified a highly dynamic pattern of HIV replication and target cell turnover, where the estimated virion half life is approximately 2 days, producing up to 10^9 new virions per day and necessitating the concomitant production of $1–2 \times 10^9$ new CD4 T-cells each day, even during the clinically latent period. In addition to the high turnover, the HIV specific reverse transcriptase is highly error prone and generates large numbers of genetically altered virions with each replication cycle, producing a 'quasi species' of genetically and phenotypically different but related isolates within the same individual.

Ultimately infection with HIV-1 leads to destruction of the immune system with a gradual decline in total numbers of circulating CD4 T-cells in addition to structural destruction of the lymph node architecture, crucial for antigen priming. Infection with HIV-1 has a protracted clinical course, where a long, but variable length of asymptomatic infection is followed by a steep rise in plasma viral load accompanied by a corresponding fall in CD4+ T-cell numbers with accompanying immunosuppression and infection with potentially fatal opportunistic infections.

Infection with HIV-1 instigates a vigorous immune response. The initial immunological control of HIV viremia is thought to be established by HIV specific cytotoxic lymphocytes (CTL), whilst sustained maintenance of low viremia is mediated probably by macrophages, dendritic cells (DC) and CD4 T-cell help inducing local cytokines, and neutralizing antibodies.

Currently available combination chemotherapy for HIV, whilst providing clinical stability and allowing immune reconstitution, are merely suppressive measures that are susceptible to viral resistance and severe side effects which together may jeopardize long term drug treatment. In addition, the suppressive action of combination therapy on HIV is, to date, unable to completely eradicate latent viral reservoirs— consequently cessation of treatment allows reappearance of virus. Hence much emphasis has been placed on the development of new constructs with minimal side effects that have potent anti-viral activity. In fact one of the potential advantages of gene therapy over current antiretroviral treatments may be the potential capacity to attack the viral reservoirs or block any reactivation from this quiescent pool.

7.2.3 Anti-HIV gene therapy

HIV-1 has at least 10 functional genes that are potentially susceptible to genetic manipulation; *Figure 7.4* illustrates a summary of the various HIV-1 gene therapy targets described to date, and each will be considered in more detail.

7.2.4 Prevention of viral entry into target cells

Prevention of entry into target cells can be considered to include all steps prior to insertion of viral DNA into the host cell genome. CD4 represents the key surface molecule involved in HIV attachment and entry into its target cell. It was therefore logical initially to use various techniques that could either block surface expression of CD4 thereby preventing HIV attachment to its target, or down regulating CD4 expression. Several early studies employed soluble recombinant CD4 in an attempt to bind and block viral gp120 from interaction at the target cell surface. However, the majority of these studies both *in vitro* and *in vivo*, had disappointing results. This was probably a consequence of both the variability of the isolates of HIV-1 themselves which have been shown to have different 'neutralization' properties in terms of CD4, in addition to poorly tolerated side effects of immunosuppression, a result of the functional role of CD4, which has a vital immunological activation role.

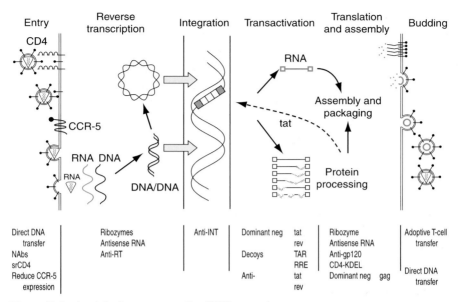

Figure 7.4. Antiviral constructs for HIV gene therapy.

Although CD4 is vital for HIV entry into a target cell, the chemokine co-receptors, which appear to have less generalized functions, also represent a pivotal role in allowing viral–host cell fusion. One per cent of Caucasians naturally express the HIV protective 32 base pair deletion of the CCR5 chemokine receptors (delta-Δ32) and fail to express any surface CCR5 and yet are otherwise immunologically competent. This has led to the use of CCR5 as a target for gene therapy. Retention signals have been incorporated into the carboxy terminus of the CCR5, CXCR4, SDF-1 receptors called 'intrakines'. These modified intrakines, so called because they are retained within the cell, bind to the cognate receptors and trap them within the endoplasmic reticulum (ER) where they are rapidly degraded. The safety of these manipulations have to an extent been substantiated *in vitro*, where the Δ32 CCR5 modified lymphocytes appear to function normally. Such intrakine gene therapies offer further advantages over new therapies involving direct administration of bioactive chemokines; first, they circumvent the problem of short half lives (< 10 minutes *in vivo*). Second, they minimize any potential in-flammatory side effects associated with giving high doses of inflammatory mediators. Third, in a similar way to CD4, all strains of HIV bind to either CCR5 or CXCR4, and therefore strain specificity is also to an extent overcome. Furthermore, the predominant strain of HIV trans-mitted sexually has been shown to utilize CCR5 as its co-receptor and is macrophage tropic, therefore if host mucosal macrophages are transfected with such an intrakine construct prevention of HIV transmission in an uninfected host could be a real possibility. The strategies that prevent viral entry are more effective at reducing overall viral burden than techniques that simply reduce viral replication.

Pre-integration gene therapy for HIV. Following viral entry into its target cell the viral RNA is reverse transcribed into DNA, a form capable of integrating into the host cell genome. In order to prevent the formation of HIV DNA various genetic manipulations have been attempted which act to block the process of reverse transcription. Whilst several antiretroviral drugs now available act to specifically block HIV reverse transcriptase (nucleoside and nonnucleoside reverse transcriptase inhibitors, RTIs) gene therapy reagents have been generated that also target this virus-specific enzyme. The majority of these techniques employ antisense RNA or enzymatic ribozymes that cleave generated nucleic acids. *Figure 7.5* illustrates the events that have been targeted.

7.2.5 Antiviral constructs

Antisense RNA. *Figure 7.5* illustrates how anti-sense RNA in conjunction with ribozymes act both pre- and post-integration to block new virion integration and formation, respectively. In comparison with protein inhibitors, RNA-based techniques are less likely to be immunogenic, less likely to interfere with normal cellular functions and are easier to express at higher levels.

The principle employed with these genetic constructs utilizes the specific Watson–Crick base pairing to interfere with gene expression in a sequence specific fashion. The engineered anti-sense RNA construct is targeted to bind to a specific region of viral RNA thereby preventing its reading and blocking the production of its product, DNA, hence preventing integration into host cell DNA. Several studies have investigated the application of stable expression of viral antisense RNA that is capable of inhibiting HIV replication. Initial descriptions employed anti-sense RNA for regions of the HIV specific LTR resulting in significant reduction in viral production. Transfection of an antisense LTR construct into human CD4+ T-cell lines resulted in a three log reduction in viral titer compared to control cells following viral challenge *in vitro*. Initial transcripts to *tat, rev, vpu, gag* and primer binding sites have shown limited efficacy. However, outside HIV infection more successful results have been obtained studying HTLV-1, MMLV and CMV [2].

Ribozymes. Anti-sense RNA alone is often not sufficiently potent to overcome potential viral resistance and in order to optimize treatment, ribozymes are included which digest the pre-targeted fragments of RNA. Ribozymes are enzymatic RNA molecules that can specifically recognize and cleave target RNAs, and prevent or significantly reduce translation of

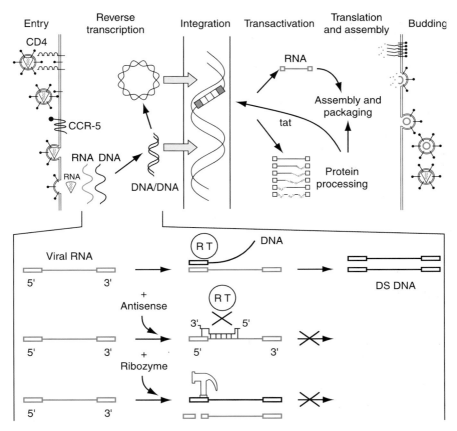

Figure 7.5. Antiviral constructs: ribozymes and anti-sense RNA.

proteins encoded by the sequences against which they are targeted. (see *Figure 7.5*).

Hammerhead ribozymes are RNA molecules that hybridize to complementary RNA sequences where the central portion of the sequence forms a specific secondary structure. Reactive groups are located within this secondary structure that mediate specific cleavage of the target RNA at a consensus target. Ribozymes specific to HIV conserved genes such as *tat rev gag*, *vif* and *integrase*, have been demonstrated to be highly efficient at reducing viral protein expression. Studies using a 'hairpin ribozyme' designated to specifically cleave the 5′ leader sequences of HIV RNA have also demonstrated long-term resistance to challenge with clinical isolates of HIV in primary human T-cell lines *in vitro*.

Ribozymes have several advantages for use as gene therapy constructs; first, because of their catalytic capacity they are effective at lower concentrations, allowing preserved efficacy at lower levels of transduction efficiency. In addition the high fidelity and substrate specificity reduces potential toxicity. Ribozymes can also be engineered to target both the pre- and post-integration phases of viral replication by cleaving incoming virion RNA in addition to transcribed and sub-genomic mRNA. Finally, they are nonimmunogenic. Clinical trials of such ribozyme gene therapies have optimistic outcomes. However, there are two potential problems with the use of ribozymes alone; first, their function is highly dependent on exact concentrations of substrate, and there may be anticipated reduced efficiency where the concentration of substrate available is limiting. In the clinical setting, where gene therapy is envisaged as additional therapy following antiretroviral treatment, and where plasma viral load is maintained at < 50 copies per ml of blood, limited HIV substrate might reduce efficacy. Second, the induction of viral resistance to the site of RNA cleavage would prevent ribozyme mediated digestion. This has recently been circumvented by the cotransduction of multiple sets of ribozymes complementary for different viral RNA sites. More recently CCR5 specific ribozymes have been developed that appear to effectively cleave their target *in vitro* and can be expressed to measurable levels in target cell lines. Here ribozyme modulated reduction of surface expression of CCR5 will prevent viral entry into uninfected cells.

7.2.6 Post-integration antiviral gene therapy

Once inserted into the host cell genome the HIV DNA remains quiescent whilst the host cell remains latent. This pool of HIV represents a small but long lived reservoir of infection and is extremely difficult to target using conventional therapeutic agents [3]. The advantage of gene therapy techniques over conventional therapies for controlling these latent HIV reservoirs is that once successfully transduced the gene therapy reagents will also remain dormant for the entire life span of the HIV infected cell and will potentially block replication and new virion formation whenever it should occur, even if that is many years after treatment is completed. One technique employed to block HIV-1 RNA transcription in latently infected cells, is that of decoy RNA.

RNA decoys. RNA decoy strategies have been used which act to inhibit viral RNA production and/or function. The advantage of RNA decoys over protein inhibitors is they are less likely to initiate an immune response. RNA decoys employ short RNA oligonucleotides which mimic critical regulatory sequences in HIV. The majority of the RNA decoys have targeted the HIV regulatory gene products *Tat* and *Rev*. Decoys such as TAR and RRE inhibit the action of the *Tat* and *Rev* regulatory sequences, respectively, by sequestrating the RNA binding proteins (see *Figure 7.6*). Several studies have shown the inhibition of HIV replication by the expression of either TAR or RRE decoy sequences. Here the host cell divides following transactivation replicating viral DNA in the process. *Figure 7.6* demonstrates the site of action of the RNA decoy strategies for

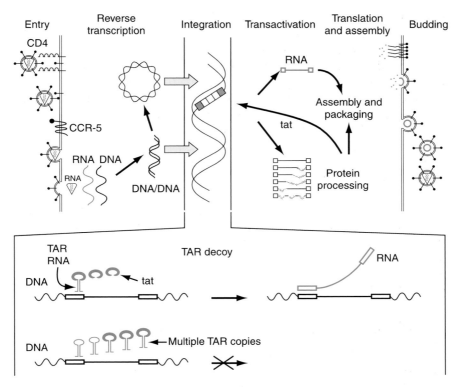

Figure 7.6. Antiviral constructs: decoy RNA.

both *Tat* and *Rev* gene products. Where, in the presence of an RNA decoy the viral DNA synthesis is interrupted, thereby preventing formation of a complete RNA molecule and hence blocking new virion formation. However, as with all gene therapy techniques that rely on post-integrational events, the viral genome remains within the host cell genome, and therefore should the level of construct be limiting or in some way sub-optimal it is possible that natural viral replication may resume.

Dominant negative mutants: transdominant proteins. Dominant mutations may be constructed to interfere with viral life cycles (*Figure 7.7*). For example, Freidman and colleagues demonstrated that genetically modified herpes virus proteins interfered with the replication of a wild type virus, and halted disease progression [4]. In this study, it was shown that a truncated form of the herpes virus transactivator sequence protein VP16 was capable of preventing wild type viral infection and replication. The wild type VP16 activates transcription of HSV 'immediate early' genes and sets in motion the entire viral transcription process. The truncated fragment of VP16, which lacks a part of the protein capable of activating transcription, competes with the protein made by an infecting virion and

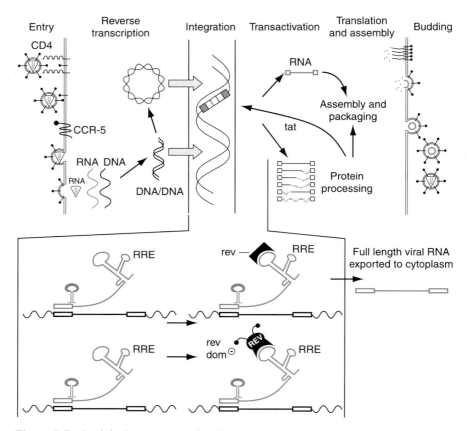

Figure 7.7. Antiviral constructs: dominant negative mutants.

thereby prevents the new virus from activating transcription of its own genes. In this way, cells have been genetically engineered to become resistant to viral infection and replication.

Following the success of experiments using mutant CMV proteins preventing viral infection of target cells, a similar application has been proposed for the HIV-1 regulatory proteins; *tat* and *rev* [5], in addition to the structural proteins; *gag* and *env*. Subsequent to DNA transactivation the viral RNA is synthesized however, export of full length viral RNA from the nucleus to the cytoplasm in preparation for assembly and viral budding is controlled by a regulatory protein product *rev*. The *rev* protein binds to an RNA sequence, the *rev*-reactive element (RRE). In an identical fashion, the genetic construct of a dominant negative mutant *rev* binds specifically to the naturally occurring RRE ligand, but following binding fails to allow transport of the full length RNA into the cytoplasm thereby blocking assembly and new infectious virion formation.

A transdominant *rev* protein has been studied most extensively, where mutations of a functionally vital well conserved leucine rich region close to the C terminus has been shown to yield defective proteins that act as transdominant inhibitors of wild type *rev*. In addition, it has been shown that mutant HIV *gag* gene products, which ordinarily exist in highly multimerized forms, were also capable of interfering with the formation of infectious viral progeny. A similar transdominant *env* protein, has also been shown to have potent effects on viral infectivity by interfering with the CD4-gp120 interaction of wild type virus with its target cell.

There are, however, several disadvantages of transdominant proteins as a form of gene therapy for infectious diseases. Firstly, since their mode of action is via suppression of virion maturation, to be effective they are confined to post integration events; unless the construct is 100% efficacious, there will remain potentially viable integrated viral DNA. In addition the expression of viral proteins synthesized endogenously by the infected target cell will be anticipated to initiate an immune response in their own right, ultimately resulting in potential destruction of the vector expressing cell. For these two reasons other inhibitory genes that target viral replication at other sites may be preferable.

7.2.7 Sequestration of viral proteins

Throughout the entire viral life cycle there are many potentially susceptible proteins that can act as targets for antibodies generated *in vitro*, see *Figure 7.8*. In order to prevent viral entry neutralizing antibodies, which have been isolated from humans naturally infected with HIV, do appear to prevent infection of cells *in vitro* at high enough concentrations. However, the majority of these neutralizing antibodies are viral strain specific, since they are directed against the highly variable areas within gp120, the outer envelope glycoprotein of HIV-1. The main disadvantage of the *env* gene product as target for anti-gp120 therapy is the huge variation in tertiary structure between isolates of HIV-1, and in view of the lack of cross reactivity between isolates of HIV-1 will not represent a feasible option for the gene therapy of HIV infection.

Within an infected cell, viral protein synthesis employs the ribosomes and endoplasmic reticulum of the host cell, and several retention signals have been previously demonstrated to 'trap' the synthesized protein within the endoplasmic reticulum (ER) permanently, preventing its incorporation into the complete virion and thereby blocking new virus formation. Several different groups have applied this principle to HIV using a tetrapeptide ER retention signal; KDEL sequence (see *Figure 7.9*). The KDEL signal was first incorporated into soluble CD4 (sCD4-KDEL), and this complex was then co-transfected into a target cell along with the HIV-1 envelope glycoprotein, gp120. Gp120 binds to CD4 with high affinity, and therefore gp120 when synthesized, binds to the transfected sCD4-KDEL and remains trapped within the ER, and failing to permit new virion formation. All strains of HIV-1 gp120 will bind CD4, and

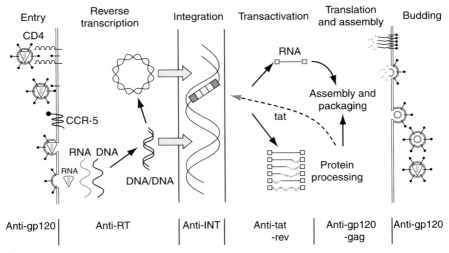

Figure 7.8. Antiviral constructs: intracellular antibodies.

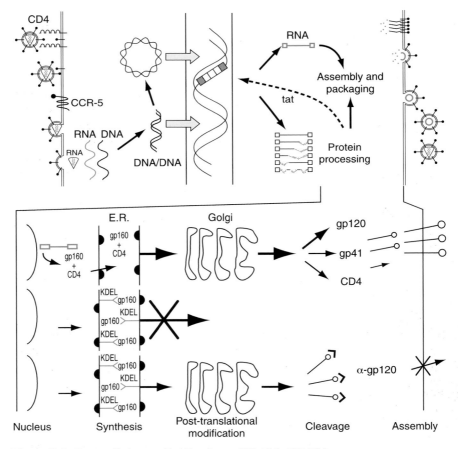

Figure 7.9. Intracellular antibodies (e.g. sFV 105, KDEL).

hence a unique advantage of this technique is to overcome the problems of strain specificity encountered with gp120 specific antibodies previously discussed.

A similar technique has been described using intracellular antibodies in the place of the KDEL retention signal. This group has designed a single chain antibody specific for gp120 that retained newly synthesized gp120 within the ER. In this study cells isolated from HIV infected donors were co-transfected with a single chain antibody directed against the envelope sequence of HIV-1. Because the single chain immunoglobulin is not transported without its partner chain from the ER into the cytoplasm it anchors the newly synthesized gp120 within the ER. However, the problems encountered with differences between laboratory isolates of HIV and primary isolates in addition to the highly variable sequences of gp120 have failed to circumvent HIV strain specificity, and therefore failed to provide clinical viability of this therapy. Single chain immunoglobulin models directed against *Rev* that retains the *rev* gene product within the cytoplasm have also been described.

7.2.8 Summary of antiviral constructs used for the gene therapy of HIV-1 infection

To summarize, the main antiviral gene therapy techniques employed for the treatment of HIV infected individuals would probably be used in combination with triple antiretroviral chemotherapy which will act to significantly reduce the overall burden of HIV infected cells. The function of the gene therapy construct should be to suppress the key viral reservoirs that form the pool of new infectious virions and represent the probable source of drug failure. It is likely that combination antiretroviral therapy will ultimately fail in many or most individuals because of poor compliance, intolerable side effects and the ever present pool of replication competent virus. Thus immunological mechanisms are also being pursued.

7.3 The use of gene therapy to manipulate the immune response

Two forms of immunological manipulation are capable of preventing disease: (a) to induce artificial immunological memory capable of preventing established infection ever occurring; prophylactic vaccination; and (b) to enhance the naturally occurring, ineffective immune response, following established infection thereby clearing the pathogen; therapeutic vaccination.

7.3.1 Understanding the immune responses

In order to appreciate the potential role that genetic manipulation of an immune response may have, a clear understanding of the mechanisms underlying the induction and maintenance of an efficient immune response is necessary.

The immune system has to fulfil two main roles: first, to recognise 'non-self' foreign protein challenge, and second, to remove the invading pathogen via a complex interplay of effector systems. Generation of such an immune response requires integration between all arms of the immune system, and the development of immunological memory to prevent re-infection. 'Self' is recognised by the expression of the same major histocompatability complex (MHC) proteins expressed by all nucleated cells within an individual, and foreign proteins can only be recognised in association with 'self' MHC molecules. *Table 7.1* lists the different effector functions of the cells of the immune system.

Table 7.1. Effector cells of the immune system

Effector cells	Effector function	Principal infectious targets
Plasma cells	Immunoglobulin secretion	Bacterial infections
Cytotoxic T cells (CTL)	Target cell lysis	Viral infections
Helper T cells	Cytokine secretion	Potentate all immune responses especially viral/parasitic
Natural killer (NK) cells	Antibody dependent cell mediated cytotoxicity (ADCC)	Intracellular pathogens, virus
Macrophage	Phagocytosis/microbicidal	Intracellular organisms, bacteria, parasites (TB)
Neutrophil	Phagocytosis/microbicidal	Intracellular organisms, bacterial infections
Eosinophil	ADCC	Helminths/parasitic
Basophil	Release of vascular permeability factors	Bacteria
Mast cells	Release of vascular	Bacteria/parasites

7.3.2 *Effector cells of the immune system*

The effector cells of the immune response originate in the bone marrow where they develop from pluripotential hemopoietic stem cells, termed progenitor cells. Immature antigen 'naive' cells leave the bone marrow, and circulate in the blood and lymphatic systems to specialized organs such as lymph nodes, and specific areas within all tissues such as gut, lung, brain, heart, kidneys, mucosae and skin. Upon challenge with appropriate anti-gen these cells mature, differentiate and develop their effector functions, a proportion of these effector cells remain as hyper-responsive memory cells, with retained specificity towards a particular antigen. *Table 7.1* illustrates the different cells involved in the generation of an immune response. Although naturally occurring immune responses generally incorporate all arms of the immune system certain cell types preferentially clear certain types of infections, a function of the methods employed by the invading organism to gain entry and reside within host cells. It is clear that for the majority of bacterial infections the principal role for clearance is via a specific antibody response, in addition to neutrophils and phagocytosis. Whilst for viral infections cytotoxic CD8 T-cell responses effect most of the pathogen clearance in addition to phagocytes, and NK cells. For helminth and parasitic infections eosinophils and antibody responses predominate.

7.3.3 Antigen recognition

In order to generate an immune response the foreign protein must be recognized as 'foreign', this is achieved by the expression of pathogen protein fragments by infected target or antigen presenting cells in association with self MHC. The exact characteristics of the foreign peptide fragment or 'epitope' seen by effector cells, is a function of both the forces that control binding into the peptide binding groove of self MHC in addition to the structure and conformation of the peptide itself. This protein fragmentation process (antigen processing) occurs in all cells and generates small (9–20mer) peptides, which are presented in association with either MHC class I (for CTL) or class II (for CD4 T-cell help). Which pathway the foreign protein enters is dependent on the route of protein entry into the host target cell. The different pathways are illustrated in *Figure 7.10*, indicating that proteins synthesized endogenously are generally expressed via the MHC class I route, (*Figure 7.10a*) whilst proteins taken up from the cytoplasm or extracellular space are processed for association with MHC class II, (*Figure 7.10b*). In order for a vaccine to mimic the immune response initiated during natural infection with an organism the vaccine proteins chosen or engineered must generate identical epitopes to those seen during the course of established infection.

7.3.4 Gene therapy as protective vaccines enhancing protective responses

Historically vaccines have enhanced the protective immune responses, beginning with the use of cowpox to prevent smallpox as demonstrated by Edward Jenner. Here cross recognition of key immunogenic antigens expressed in the harmless infectious agent that causes cowpox, provided protection, by initiating immunological memory, against the smallpox virus, and thereby prevented establishment of smallpox infection following exposure. This was followed by the adoption of a similar principle whereby virulent organisms were in some way altered via passage *in vitro* (live attenuated) or via chemical or heat inactivation processes (fixed inactivated) which rendered them nonpathogenic whilst retaining their immunogenic properties. *Table 7.2* summarizes from a

(a)

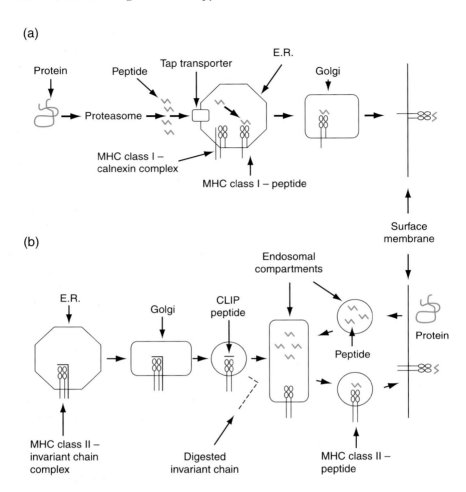

Figure 7.10. The processing and presentation of antigen. (a) MHC class I pathway. (b) MHC class II pathway.

Table 7.2. Summary of current vaccinations

Traditional vaccines		New vaccines			Gene vaccines
Fixed inactivated	Live attenuated	Recombinant DNA protein	Peptide	Viral vectors	Naked DNA
Salk polio vaccine	Sabin polio vaccine	HBV sAg	HIV-1 gp120 subunit	Vaccinia MVA	Influenza A
Problems					
Safety Immunogenicity	Safety Loss of attenuation		Defining protective epitopes	Safety	Safety

historical perspective different techniques employed for vaccination. Fixed inactivated vaccines have the advantage of being safe and generally effective, but the disadvantage that with the fixing process key immunogenicity is impaired providing a suboptimal immunological response that reduces efficacy. In an attempt to retain immunogenicity and efficacy certain pathogens are attenuated *in vitro* by a variety of techniques which renders them nonpathogenic or in some other way alters their pathogenicity but maintains replication. These live attenuated vaccines have the advantage of retaining more of the naturally occurring immunogenic properties, but the disadvantage being the potential for attenuated organisms to revert back to pathogenic phenotype. Examples of live attenuated vaccines include: Sabin polio; measles; mumps; rubella; and yellow fever. Both live attenuated and fixed inactivated vaccines contain many antigens not required for protection, and hence have potential unwanted side effects.

Newer vaccines employ genetic manipulation of parts of the organism, chosen for their immunogenic capacity. Such vaccines employ recombinant DNA proteins, an example of which is the hepatitis B surface antigen vaccine. The main problem of recombinant protein vaccines is their type specificity. There is the small but real incidence of mutant strains of the wild type virus that have altered surface antigen proteins such that they are not cross recognized by the immunological priming of the recombinant antigen. Additionally, the production of large volumes of purified proteins have proven practically difficult. Peptide vaccines have also been tried, but the main problem in generating new peptide vaccines has been the exact mapping of relevant epitopes enabling homing down from an entire invading pathogen to a short peptide sequence, that withstands all MHC classes.

One problem encountered with synthetic protein vaccines has been an inability to replicate the exact epitopes generated in a natural infection. Hence whilst retaining the capacity to generate measurable immune responses *in vitro*, synthetic protein or peptide based vaccines have been unable to provide protection against natural exposure and infection. Such is the case for HIV-1 infection, where numerous vaccine preparations using the HIV outer envelope protein have generated neutralizing antibody titers *in vitro*, in addition to CTL and CD4 activity but have failed to provide sterilizing immunity when challenged with field isolates of virus in animal models

The use of gene therapy to insert specific viral DNA that allows *de novo* intracellular synthesis of viral proteins and will more accurately mimic natural infection should therefore generate a more appropriate immune response. Such immunization by *in vivo* expression of viral proteins for prophylactic vaccines, or for modulation of the immune response in the face of existing infection has been demonstrated for certain infectious diseases in animal models; for example using an influenza A virus murine model. These studies employed direct intramuscular injection of saline containing only pure plasmid encoding the influenza A nucleoprotein. Using this technique it was demonstrated that persistent gene expression of the influenza protein was possible, and in addition the vaccine was able to generate measurable influenza specific CTL activity, capable of providing demonstrable protection from an otherwise lethal challenge with the influenza virus in the mice. Influenza A is a highly variable virus, capable of inducing only strain-specific immunity following natural infection, or after uptake of the currently available vaccines. The data cited above demonstrated that mice receiving the DNA vaccine were resistant to viral challenge with heterologous strains arising 34 years after the vaccine strain, a significant advantage over current vaccines. For practical uses of such technology conserved regions of the organisms must be chosen in preference to more variable proteins such as the envelopes, thereby circumventing the strain specificity limitations of the currently available vaccines. This and other reports have suggested that such 'naked DNA' vaccines capable of generating naturally occurring epitopes may offer the ability to mimic the immunologically critical aspects of natural infection necessary to generate sterilizing immunity and thereby afford protection against natural infection. Similar applications have been studied using direct DNA injection of the HIV-1 envelope glycoprotein, gp120. Use of these gene therapy DNA vaccinations have generated measurable HIV-1 strain specific neutralizing antibodies as well as the induction of specific lymphocytic responses. However, many have failed to protect against *in vitro* challenge with primary isolates of HIV-1, which are known to differ in the tertiary conformation of the gp120 molecule, in addition naked DNA vaccines have failed to induce a sustained immune response. Other studies using murine models and retroviral delivery systems based upon gp120 envelope proteins have demonstrated similarly effective neutralizing antibody and specific CTL activity in recipient animals, however, cross-strain protection has not been demonstrated using these techniques. Using animal models, genetic immunization by direct DNA vector has been shown to effectively induce humoral and cellular immune responses against HIV-1 using *env* antigen [6] and in non-human primates using *env* and *rev* antigens [7,8]. Studies of chimpanzees similarly immunized have demonstrated protection following challenge with HIV-1 indicating effective immunization [9].

An additional advantage of this system is to avoid the necessity of high level transgene expression, in fact low level antigen expression has been shown to be advantageous in the induction of the optimal Th1 type immune responses shown to be associated with effective viral clearance [10].

More recently a novel combination of DNA vaccination followed by boosters using vectors coupled with specific HIV epitopes have generated effective and sustained protection in animal models [11]. Based on the principle that for an effective vaccine against HIV there must be efficient induction of an HIV specific CTL response, multi-cytotoxic T-lymphocyte epitope antigens were employed in combination with nucleic acid technology and were shown in murine models to induce effective HIV specific CTL responses [12] which were protective. To maintain a sustained response the DNA vaccine was followed up using a single boost with a recombinant modified vaccinia virus Ankara (MVA). Using this technique high level sustained antigen specific CTL were generated against both HIV and malaria. In the malaria study, a particular sequence of subunit immunization with pre-erythrocytic antigens of *Plasmodium berghei*, consisting of single dose priming with plasmid DNA was given. This was followed by a single boost with a recombinant modified vaccinia virus Ankara (MVA) expressing the same antigen and induced unprecedented complete protection against *P. berghei* sporozoite challenge in two strains of mice [13]. Using a similar technique high level protection has also been demonstrated with HIV, where poly-epitope genes were directly transferred using nucleic acid technology, followed by a single MVA booster expressing the same epitopes. These vaccines were shown to induce HIV specific interferon gamma producing and cytolytic CD8 + T-cells after single vaccine administration in mouse models [14]. These techniques are to be introduced as a phase I trial for HIV vaccination programs for humans.

7.3.5 Therapeutic vaccines – postinfection immunotherapy

The aim of therapeutic vaccines is to enhance naturally occurring immune clearance of infection. Only rarely have certain infectious diseases been targeted in this way. In general following infection with any organism in the absence of pre-existing immunosuppression the immune response naturally generated is optimal and very little enhancement capacity is possible. The exception to this has been the relative success of the post-exposure vaccination against rabies virus, and hepatitis B. Under such circumstances the balance between exposing dose of virus and timing following exposure to vaccination is critical and may account for those cases of clinical failure.

Unfortunately, the use of therapeutic vaccines for the treatment of HIV infection has been to date inconclusive. The strain-specific immunity generated from a gp120 specific vector delivery system, whilst enhancing any existing HIV specific immunity may have little or no influence on replication and infectivity of the infecting strain within the host. In addition, all clinical stages of HIV infection are associated with marked CD4 T-cell dysfunction, often disproportionate to the total T-cell count [15]. In addition, asymptomatic HIV-infected individuals also have impaired capacity to expand CTLs *in vivo* following priming [16]. Hence in the context of pre-existing immunosuppression the efficacy of a therapeutic vaccine aimed at clearing established HIV infection will be sub-optimal. Furthermore, there is no paucity of expression of viral proteins within an HIV-infected host even in early stages of infection, which despite inducing a vigorous humoral and cellular immune response fail to control viral replication or lead to viral eradication.

7.3.6 Adoptive transfer of genetically altered immune cells

Although the possibility of therapeutic vaccines controlling viral disease in HIV seem remote, the adoptive transfer of genetically altered autologous effector CTL may enhance viral clearance. In addition, infections that cause problems for immunosuppressed hosts, such as following bone marrow or organ transplantation are also susceptible to such therapy.

A model of adoptive transfer of genetically altered cytomegalovirus (CMV) specific CTL clones was first described in mice. CMV represents a major health problem for the immunosuppressed patient, irrespective of the cause of their immunosuppression and CMV specific immune enhancement would have a clear clinical application. In this study CMV specific CTL developed during primary infection isolated and expanded *in vitro* and then re-introduced into the same mouse. This technique was able to confirm protection against viral challenge, even in the face of established CMV pneumonia. A similar application of this principal to humans has similarly demonstrated persistent reconstitution of a CMV specific CTL response [17] but to date has not conclusively been shown to either prevent disease or enhance recovery from established infection with CMV.

In addition, one theory proposed to explain the ultimate failure of the immune system to clear infections such as HIV is based on CTL 'exhaustion'. Here, the overwhelming antigenic challenge observed during HIV infection leads to massive and inappropriate clonal expansion of antigen specific T-cells, and, in a manner analogous to the hypothesis whereby superantigens lead to clonal cell death, thereby deleting an HIV specific repertoire from the immune response. If, however, this HIV specific clone was artificially enhanced then viral clearance may be possible. The ability to perform similar manipulations of HIV infected cells is uncertain. In addition preliminary attempts to reconstitute HIV infected patients with artificially expanded HIV specific CTL had unfortunate results. In these cases, the few human studies performed showed that the HIV specific CTL contrary to expectation, caused an aggressive fall in CD4+ T-cell numbers with subsequent clinical deterioration, assumed to be a result of over enhanced CTL removal of all HIV expressing T-cells. In an attempt to overcome such immune mediated cellular destruction HIV specific CTL were similarly transfected with a vector containing the Hygromycin resistance gene and a 'suicide gene'; a hybrid of the herpes simplex virus (HSV) thymidine kinase gene (HSVTK). This manipulation allowed specific selection of transduced HIV-specific CTL using Hygromycin containing medium, which were then transferred back to the patients. The addition of the HSVTK circumvented the potential cell death seen in previous studies whereby, in the event of similar overwhelming target cell destruction the administration of the anti-HSV drug, Gancyclovir, resulting in HSVTK catalyzed phosphorylation would provide selective killing of the gene altered cells. Similar HIV specific gene therapy is currently being investigated, where the advantage of the potential suicide gene allows greater safety.

7.3.7 Cellular immune responses to gene therapy constructs

One major potential obstacle to gene therapy is the elimination of such gene-modified cells by the immune system reacting to the novel protein products of the introduced genes. Since some of the earliest attempts at gene transfer it has been recognised that specific B- and T-cell responses are elicited against components of viral vectors that limit gene expression [18]. In previous studies, recombinant adenovirus proteins have been shown to initiate a potent immune response thereby limiting the persistence of transduced cells and preventing successful gene delivery [19]. One method of overcoming vector specific immune responses initially tried was to use gene therapy in combination with immuno-suppressive strategies such as cyclosporin A, however these often had systemic immuno-suppressive side effects. In addition, construct specific immunity has been previously demonstrated to occur *in vivo*, in the context of host immunosuppression with HIV-1 infection. More recently a comparison of the adoptive transfer of autologous unmodified and gene modified CD8 + T-cells directed against certain CMV proteins to an HIV immunosuppressed host in an attempt to enhance any existing anti-CMV effect clearly demonstrated the induction of T-cell immune responses specific to the autologous cells modified by retrovirus mediated gene transfer [20]. The type of gene therapy target cell has some bearing on the nature and extent of the immune responses initiated, in addition to the type of construct employed. If antigen presenting cells are the host cell targets, as is the case for naked DNA vaccines, it is logical that a more potent anti-construct immune response will be generated. In addition, post-integration anti-viral constructs that consist of expressed proteins are also more likely to induce an immune response than those utilizing RNA. These data emphasize that successful gene therapy will require the development of strategies to allow the novel proteins expressed in gene modified cells to be less immunogenic.

7.3.8 Summary of gene therapy manipulation of the immune system

There is little clinical experience to offer promising outcomes for any of the immuno-gene therapy techniques addressed above. *Table 7.3* illustrates a summary of the mechanisms employed in HIV specific immuno-gene therapy. The probability that enhancement of the naturally occurring fully stimulated immune response by the presence of vast

Table 7.3. Summary of HIV-1 specific immuno-gene therapy

Mechanisms of action	Examples
Enhance existing immune responses	Over-expression of gag CTL epitopes
Expression of cytokine/chemokines	CCR-5 production
Exogenous clonal expansion	HIV-1 specific CTL expansion

amounts of viral antigens throughout the course of HIV-1 holds little hope for an effective therapeutic vaccine using this approach. Although passive transfer of autologous expanded CTL clones may offer temporary suppressive action. There is more optimism regarding increased expression or secretion of the inhibitory CTL derived factors in particular, in addition to augmentation of the chemokine proteins, either locally or systemically. At the site of HIV transmission (the genital mucosae) gene therapy constructs could be delivered and act to alter local concentrations of chemokines or specifically downregulate the surface expression of the chemokine co-receptors for HIV thereby preventing viral attachment, entry and sexual transmission.

7.4 Conclusions

The wealth of research that has been invested into potential gene therapy for infection with HIV-1 in terms of both antiviral and immunomodulatory therapy, can be employed successfully in the future for use alongside conventional treatments. Using HIV as an infection model, the technology may also be applied to many other suitable infectious diseases. Such opportunities can be usefully expanded to other retroviral infections such as HTLV-1 and HIV-2, in addition to the herpes group of viruses which cause significant mortality in the immunosuppressed host and neonate, EBV, CMV and HSV, and world-wide the hepatitis viruses; B,C and delta for which there is currently no curative therapy available. It seems certain that within the next decade gene therapy for both the treatment and prevention of infectious diseases will be an encouraging much needed alternative therapy.

References

1. Paxton, W.A., Martin, S.R., Tse, D. *et al.* (1996) Relative resistance to HIV-1 infection of CD4 lymphocytes from persons who remain uninfected despite multiple high risk sexual exposures. *Nat. Med.*, **2**, 412–417.
2. Stein, C.A. and Cheng, Y.C. (1993) Antisense oligonucleotides as therapeutic agents...is the bullet really magic? *Science*, **261**, 1004–1012.
3. Chun, T.W., Carruth, L., Finzi, D., *et al.* (1997) Quantification of latent reservoirs and total body virla load in HIV-1 infection. *Nature*, **387**, 183–187.
4. Freidman, A.D., Triezenberg, S.J., and McKnight, S.L. (1988) Expression of a truncated viral trans-activator selectively impedes lytic infection by its cognate virus. *Nature*, **335**, 452–454.

5. Baltimore, D. (1988) Gene therapy. Intracellular immunisation. *Nature*, **335**, 395–396.
6. Wang, B., Ugen, K.E., Srikantan, V., *et al.* (1993) Gene inoculation generates immune responses against human immunodeficiency virus type 1. *Proc. Natl Acad. Sci. USA*, **90**, 4156–4160.
7. Warner, J.F., Anderson, C.G., Laube, L. *et al.* (1991) Induction of HIV-specific CTL and antibody responses in mice using retroviral vector-transduced cells. *Aids Res. Hum. Retrovir.*, **7**, 645–655.
8. Wang, B., Boyer, J., Srikantan, V. *et al.* (1995) Induction of humoral and cellular immune responses to the human immunodeficiency virus in non-human primates by in vivo DNA. *Virology*, **211**, 102–112.
9. Boyer, J.D., Ugen, K.E., Wang, B.A., *et al.* (1997) Protection of chimpanzees from high dose heterologous challenge by DNA vaccination. *Nat. Med.*, **3**, 526–532.
10. Salk, J., Bretcher, P.A., Salk, P.L., Clericic, M. and Shearer, G.M. (1993) A strategy for prophylactic vaccination against HIV. *Science*, **260**, 1270–1272.
11. Letvin, N.L., Montefiori, D.C., Yasutomi, Y. *et al.* (1997) Potent protective anti-HIV immune responses generated by bimodal HIV envelope DNA plus protein vaccination. *Proc. Natl Acad. Sci. USA*. **94**, 9378–9383.
12. Hanke, T., Blanchard, T.J., Schneider, J., *et al.* (1998) Immunogenecities of intravenous and intramuscular administrations of modified vaccinia virus Ankara-based mulit-CTL epitope vaccine for human immunodeficiency virus type 1 in mice. *J. Gen. Virol.*, **79**, 83–90.
13. Schneider, J., Gilbert, S.C., Blanchard, T.J., *et al.* (1998) Enhanced immunogenicity for CD8 + T-cell induction and complete protective efficacy of malaria DNA vaccination by boosting with modified vaccinia virus Ankara. *Nat. Med.*, **4**, 397–402.
14. Hanke, T., Blanchard, T.J., Schneider, J., *et al.* (1998) Enhancement of MHC class I restricted peptide-specific T-cell induction by a DNA prime/MVA boost vaccination regime. *Vaccine*, **16**, 439–445.
15. Meidema, F. (1992) Immunological abnormalities in the natural history of HIV infection: mechanisms and clinical relevance. *Immunodeficiency Rev.*, **3**, 173–193.
16. Pantaleo, G., Koenig, S., Baseler, M., Lane, H.C. and Fauci, A.S. (1990) Defective clonogenic potential of CD8 + T lymphocytes in patients with AIDS. Expansion *in vivo* of a nonclonogenic CD3 + CD8 + DR + CD25 − T-cell population. *J. Immunol.*, **144**, 1696–1704.
17. Greenberg, P.D., Reusser, P., Goodrich, J.M. and Riddell, S.R. (1991) Development of a treatment regimen for human cytomegalovirus (CMV) infection in bone marrow transplantation recipients by adoptive transfer of donor-derived CMV specific T-cell clones expanded *in vitro*. *Ann. NY Acad. Sci.*, **636**, 184–195.
18. Bromberg, J.S., Debruyne, L.A. and Qin, L. (1998) Interactions between the immune system and gene therapy vectors: bi-directional regulation of response and expression. *Adv. Immunol.*, **69**, 353.
19. Yang, Y., Nunes, F. and Berensi, K. (1994) Cellular immunity to viral antigens limits E1-detected adenovirus for gene therapy. *Proc. Natl Acad. Sci. USA*, **91**, 4407–4411.
20. Riddel, S.R., Elliott, M., Lewinsohn, D.A. *et al.* (1996) T-cell mediated rejection of gene-modified HIV specific cytotoxic T lymphocytes in HIV infected patients. *Nat. Med.*, **2**, 216–223.

Chapter 8

Targeting of gene delivery systems

Lesley-Ann Martin

8.1 Introduction

Gene therapy's simple concept of delivering a therapeutic gene to an afflicted cell is very appealing. However, the harsh reality is sobering, how do you target your gene(s) in a large enough dose to the right site and nowhere else? Targeting vectors can be achieved in a multitude of ways, with differing levels of sophistication. At one end of the spectrum, targeting can readily be achieved by simply applying the vector at its required site of action. Alternatively, genetic engineering can be used to redirect vector components to recognize, and bind, highly specific target ligands expressed on/in the target cells. This chapter addresses the physical methods of vector targeting leading on to the more complex molecular strategies.

8.2 *Ex vivo* gene transfer

Targeting by isolation is the most extreme method of physical targeting. If the disease and technology permits, target cells can be removed from the patient, cultured *in vitro* and transduced with a vector, then replaced *in vivo* (*Figure 8.1*). Such approaches are only possible where the transduced target cells are used as a source of secreted protein (e.g. to treat adenosine deaminase deficiency or hemophilia) or as a vaccine (to treat cancer).

Adenosine deaminase deficiency which causes severe combined immunodeficiency cannot be treated *in vivo* at present and as a consequence *ex vivo* targeting is the only option. All of the current protocols delivering a correct version of the ADA cDNA use retroviral

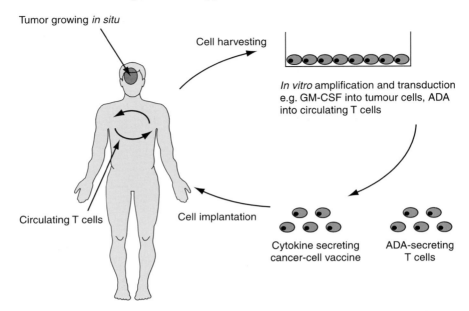

Figure 8.1. *Ex vivo* targeting. Target cells are isolated from the patient and transduced *ex vivo* with therapeutic genes, e.g. GM-CSF or ADA. The modified cells are then reinjected back into the patient. The modified cells *in vivo* then serve as cancer vaccines or machines to produce secretable protein such as ADA.

vectors to produce stable integration. Initial studies attempted to transduced hemopoetic stem cells (HSC) *ex vivo* with retroviral constructs and then to inject these modified cells. Results from mouse models were particularly successful, in fact in one study transduced bone marrow from a donor mouse was shown to maintain activity when passed onto a further recipient. However, the ability to transduce HSCs from primates and human donors proved difficult. As a consequence the most obvious choice was the T lymphocyte. In this strategy T cells were harvested, and stimulated to proliferate *ex vivo* in the presence of an ADA carrying retroviral vector, by the addition of IL-2 and a CD-3 complex antibody. Transduced cells were then reintroduced into the host. Studies showed that these modified T cells had an extended life span compared to ADA-T cells.

Hemophilia B is an X-linked recessive bleeding disorder caused by a lack of functional factor IX in the circulation. Current treatment aims to maintain the factor IX level above 4% of normal to alleviate the spontaneous bleeding into joints which can have crippling effects. The factor IX gene is expressed in hepatocytes with stringent tissue specificity. Ideally the gene should be targeted to the liver. This has however, proved difficult and other cell types have been transduced with cDNA to provide an easier target. One major requirement is that these cells used for ectopic expression can carry out all the co- and post-translational requirements necessary for correct augmentation. To this end a number of cell types have been used including keratinocytes, skin fibroblasts, bone marrow stromal cells, skeletal muscle cells and endothelial cells. The favored vectors for *ex vivo* and *in vivo* introduction of the factor IX gene have been retroviral vectors. However, these are limited to rapidly dividing cells and more recently adenoviral vectors have been used.

In the field of cancer gene therapy, investigations into the development of autologous cellular vaccines have been met with great excitement. The premise is to remove tumor cells surgically from the patient, then grow the isolated cells *in vitro*, transduce them with immunostimulatory genes, and re-inject them back into the patient in an effort to induce an immune response. Studies have shown that alteration of syngenic mouse tumor cells to secrete cytokines such as IL-2, IL-4, IL-6, TNFα, GM-CSF or γ-interferon results in immunological destruction of tumor cells *in vivo*. On the basis of these results several clinical trials have been approved, in which patients are injected with autologous or allogeneic genetically modified tumor cells. The trials involve the transduction of melanoma, colorectal, and breast carcinoma cells *in vitro* with retroviral vectors delivering the genes for IL-2, TNFα or GM-CSF. An alternative strategy is to employ autologous fibroblasts transduced with IL-2 which are then mixed with irradiated tumor cells from the patient and reinjected. The advantage of this system is that fibroblasts are easier to culture *in vitro* and also eliminates the need to culture tumor cells from large numbers of individuals.

A further strategy which exploits natural vector tropism (described in the following section) is to use tumor infiltrating lymphocytes (TILs) which have the ability to home in on tumor tissue. Several studies have attempted to exploit this property to deliver cytokines directly to the tumor mass, enhancing the T lymphocytes anti-tumor activity. However, a number of problems have been noted. First, T lymphocytes are difficult to transduce with retroviral vectors and studies suggest that the T lymphocytes downregulate the expression of cytokines from these constructs. Second, whilst many TILs localize to the tumor tissue, the majority accumulate in the liver and spleen, posing a toxic threat to the patient. Consequently application of this approach is limited and studies have focused on the use of tumor cells.

The manipulations described here are often problematic and not suitable for all diseases consequently many other targeting strategies have been developed for *in vivo* applications.

8.3 Transduction targeting

The viruses commonly used for gene therapy such as amphotropic murine leukemia virus (MuLV), adenovirus and vesicular stomatis virus recognize receptors that are ubiquitously expressed. In the following sections some of the strategies for restricting/redirecting their tropisms are discussed.

8.3.1 Exploitation of natural vector tropism

One approach to precise targeting is to exploit viruses that demonstrate a natural tissue tropism. For instance herpes viruses are ideal for delivery of therapeutic genes to neural tissue and hepatitis has the ability to target liver cells. Parvoviruses have been shown to preferentially kill transformed cells and could be used as oncoselective agents. Studies have demonstrated that coinjection of the mouse minute virus (MMV) and Erlich ascites tumor cells into the peritoneal cavity of mice inhibits tumor formation by up to 90%. Furthermore, mice that survived one coinjection were resistant to a second tumor challenge 5–6 weeks later. The mechanism of parvovirus oncotropism is unclear, but is thought to be related to an effect of the transformed cell environment on the production or activity of parvovirus autoregulatory proteins.

Mutant forms of conventional vectors are also able to demonstrate tumor cell tropism. For instance, a series of attenuated herpes viruses have been created to target glioblastomas. In the first generation of modified HSV-1 vectors the thymidine kinase gene was deleted which prevented viral replication in normal neural tissue where levels of endogenous thymidine kinase (TK) are too low for *trans* complementation. However, the level of TK in glioblastomas is significantly higher, allowing replication of the attenuated HSV-1 and cell lysis. One major problem with this system is that the attenated HSV is resistant to ganciclovir, the normal treatment for HSV infections. To overcome this a second attenuated HSV-1 was developed by deleting gp34.5 and ICP6 genes required for replication in normal neural tissue, but maintaining the TK gene and hence sensitivity to ganciclovir. This new attenuated HSV-1 can replicate exclusively in glioblastoma cells *in vitro*. Mice harboring intracranial glioblastomas, treated with this virus showed a reduction in tumor volume compared to untreated controls. The neural toxicity of vector was addressed by intracranial injection into Aotus nancymae monkeys which are sensitive to HSV-1. No encephalitis was observed. This virus is currently under investigation in a phase I clinical trial for the treatment of glioblastoma.

Vector tropism can also be exploited at other points in the transduction pathway, for instance C-type retroviruses are able to bind to cells, but are unable to infect unless the cell is undergoing mitosis. The virus requires the dissolution of the nuclear membrane for trafficking and integration of the viral genome. As such these viruses were thought to be ideal for the introduction of therapeutic genes into rapidly dividing cells in a background of quiescent nontarget cells. C-type retroviruses carrying the HSVTK suicide gene have been used to treat brain tumors. In the presence of the prodrug ganciclovir only those cells expressing the HSVTK were killed. Whilst rodent studies proved extremely successful, human studies were less convincing. It appears that human tumor cells have a much slower rate of division compared to rodent tumors with on average only 20% of cells are cycling at any one moment. Therefore this limits the efficacy of what could be a potentially powerful system.

8.3.2 Targeted integration

Nontargeted integration exhibited by retroviral vectors is potentially dangerous not only by switching on downstream oncogenes but also by direct disruption of genes. Vectors that can integrate at specific sites would eleviate these problems, such as adeno-associated viruses. These parvoviruses are naturally replication incompetent unless rescued by super-infection with adenovirus/HSV. It has been reported that adeno-associated virus (AAV) integrates in a site-specific manner on chromosome 19 at a region named AAVS1. Great interest has been shown in these vectors, first, because exogenous cDNAs driven by internal promoters can still be active in the latent integrated virus, and second, the viral inverted terminal repeat (ITR) is silent so there is no promoter interference leading to loss of tissue specificity. The harsh reality is that the packaging size of these vectors is only 4.5 kb compared to the 7 kb for retroviruses.

Attempts to target retrovirus integration have also been demonstrated by fusing the HIV-1 integrase and the DNA-binding domain of the λ-repressor. A similar result was achieved when the retroviral integrase was fused to the DNA binding protein *E. coli* LexA. Continued advances in this field are eagerly awaited.

8.3.3 Retroviral pseudotypes

The ability to produce retrovirus packaging cell lines has made the manipulation of the envelope genes far easier. This in turn has propagated great interest in the potential to pseudotype retroviruses swapping one envelope for another. It has been observed that envelope substitution occurs more readily between closely related viruses e.g. MuLV genomes can be rescued by C-type retroviruses but not D-type such as HTLV-1. Other tropisms have also been exploited such as human immunodeficiency virus (HIV) which uses the CD4 molecule for receptor mediated cell entry. Theoretically, vectors pseudotyped with the HIV gp120 envelope protein should be able to transduce CD4+ cell populations. However, problems still remain as there is often a need for additional coreceptors, also there is some suggestion that vectors pseudotyped with gp120 may contribute to the pathogenesis of HIV infection. Another study has demonstrated that a chimeric protein comprising of an in-frame fusion between the Rous sarcoma virus envelope and the influenza hemagglutinin glycoprotein can redirect the normal route of RSV cell entry. One of the most interesting strategies has been to pseudotype retroviruses by introducing the VSV G protein. The hybrid virions produced have a number of advantages over normal MuLV vectors. The virus particles can be concentrated by ultracentrifugation without disruption of the envelope and more than one copy of the virus genome can be integrated into the host genome. The mechanism by which this occurs is poorly understood but may be a result of higher viral titers in association with target cells. VSV-G-pseudotyped vectors also display a wider tissue tropism. The current development of this system is unfortunately hampered as a stable cell line expressing the VSV-G protein has not been achieved to date.

8.3.4 Targeting by retrovirus ligand conjugates

Attempts have been made to couple molecules to the retrovirus envelope to extend the host range. One approach used to improve targeting to hepatocytes has involved the chemical coupling of lactose to the ecotropic MMLV envelope glycoprotein. This process allowed the modified ecotropic envelope to recognize and specifically target the asialoglyco-protein receptor found on hepatocytes. This approach limits the vector tropism three-fold: first, to cells expressing the asialoglycoprotein receptors; second, to proliferating cells (as retroviruses can only integrate into cells undergoing mitosis); and third, as the virus is ecotropic its tropism in human cells is limited entirely to hepatocytes greatly increasing its safety compared to amphotropic retroviruses. Although this process was successful it is limited by the fact that not all ligands can be chemically coupled.

An alternative approach is to use antibodies to redirect virus targeting. In particular bifunctional antibody complexes consisting of one mono-clonal antibody directed to the retrovirus envelope joined by a streptavidin bridge to a second antibody capable of targeting a specific receptor on a nonpermissive cell. This strategy was successfully used to target the epidermal growth factor receptor (EGFR), major histocompat-ability complex (MHC) antigens (class I and II) and the transferrin receptor expressed on hepatoma cells. However, in the latter case although binding was successful, integration of the provirus was unsuccessful. It appears that not all proteins are amenable to this technology and similarly targeting *in vivo* with murine antibodies suffers from numerous disadvantages, suggesting that although this may provide a potentially powerful tool, further development will be required.

8.3.5 Engineering envelopes

Retroviral infection is iniated by the attachment of the virus envelope protein via the surface unit (SU) to specific cell receptors. The domains of the envelope (encoded by the env gene) involved in receptor binding, have been well mapped for MuLV and have become the focus for many researchers interested in modifying the tropism of these viruses. Initial studies reported the successful fusion of a gene encoding a single chain Fv fragment (directed against a cell surface antigen) to the 5′ end of the ecotropic MuLV envelope gene. The hybrid retrovirus produced was able to bind antigen mediated by the single chain Fv, but was unable to infect cells. In a similar study, fusion of a single chain Fv to the 3′ end of the spleen necrosis virus env gene produced an infectious virus able to target antigen presenting cells. Although the virus was targeted its infectivity compared to the wild-type virus was too low to be of therapeutic value. It is clear that certain sites on the SU domain can tolerate polypeptide insertion or replacement and using computer predictions it should be possible to forecast which amino acid sequences are exposed or needed for flexibility. Using this strategy, alteration of the avian leukosis viruses (ALV) tropism was achieved by inserting a 16 amino acid sequence containing an RGD motif into a predicted β-turn of the envelope protein. The hybrid virus produced was able to infect cells expressing the ALV receptor as well as mammalian cells otherwise refractory to ALV. However, insertion of longer more structured peptides (with c-loops or secondary structure) impaired envelope processing and incorporation into viral particles.

In an attempt to understand why the infection with these modified viruses was so poor, Cosset and Russell sought to determine what factors influenced infectivity. They made several MuLV-chimeric envelopes, by inserting into the N-terminal domain of the SU, ligands to target EGFR and Ram-1 [a receptor used by amphotropic murine leukemia virus (MLV-A)]. Both hybrid virions were able to recognize their respective receptors. However, the EGF chimeras were noninfectious unlike those expressing Ram-1 although only at a fraction of the efficacy of the wild-type amphotropic virus that binds Ram-1 naturally. They postulated that the non-infectivity of the EGF hybrid was a result of the EGF ligand acting as an agonist for target receptors, as a result EGFR-bound virus was taken up by lysosomes leading to abortive infections. Further studies examining the poor infectivity ratio for the Ram-1 MuLV hybrid, demonstrated the importance of chosing the correct interdomain spacing between the foreign sequence and the SU protein. Fusion of Ram-1 to codon 1 of the SU domain led to a 100-fold increase in infectivity compared to fusion at codon 6/7. It is most likely that at codon 1 the inserted ligand interferes less with correct envelope folding. Despite the problems outlined, reports of successful cell targeting where the whole receptor binding domain of MuLV was replaced, have been made. Replacement of the whole receptor binding domain by a high affinity ligand is indeed an attractive proposition. Unfortunately, interpretation of these experiments has been contested and considerable difficulty in repeating the results has been reported by a number of groups.

One of the most exciting targeting strategies to date has been the development of protease-activatable vector, which allows the amphotropic MuLV to use its natural cell receptor-mediated entry mechanism, in conjuction with a hybrid receptor (*Figure 8.2*). In this system the EGF ligand was fused to codon 1 of the SU domain upstream of a factor Xa protease cleavage site. The hybrid was able to target the EGF receptor but did not infect the cells. However, upon the addition of protease factor Xa the virus infectivity increased more than 100-fold. The mechanism allows the virus to be targeted/accumulated at specific cells via the novel receptor binding activity followed by cleavage with a site specific protease allowing virus internalization through its natural receptor route. Further modification of the system by incorporating metalloprotease cleavage sites in the env gene (both up-regulated in tumors) has produced an extremely effective targeting system for the delivery of therapeutic genes for cancer treatment.

8.3.6 Adenoviral vectors

Infection of cells by adenovirus particles takes place in two stages. Firstly the knob domain of the fiber protein binds to the coxsackie-adenovirus receptor followed by internalization mediated by the interaction between the RGD sequences encoded on the penton base and cell surface integrins. One of the major problems associated with the use of adenovirus vectors for gene therapy is the ubiquitous nature of the fiber protein receptor making targeting difficult. Attempts have been made to re-direct adeno-

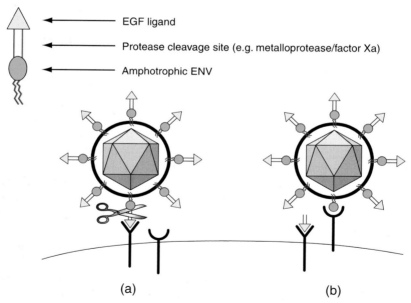

Figure 8.2. Two-step retroviral targeting strategy using protease cleavage. The EGF ligand is placed upstream of a protease cleavage site at codon 1 of the SU. (1) The EGF ligand directs binding and accumulation of the virus at the target receptor. (2) Circulating proteases then cleave the virus away from the ligand allowing the virus to enter the cell via its natural infectious route.

virus binding specificity by conjugating molecules to the fiber protein. In one study the carboxy terminus of the gastrin releasing peptide (GRP) was fused to the human adenovirus type 5 fiber protein. The fiber-GRP fusion was able to assemble into trimers with a quaternary structure similar to the wild-type fiber. More importantly the fiber-GRP was bound by anti-GRP antibodies indicating that the displayed GRP ligand was accessible to bind GRP-receptors, confering a new tropism to the adenovirus.

Most of the studies so far reported, have concentrated on the modification of the penton base which has five copies of an RGD motif (integrin receptor-binding motif) allowing interaction with the vitronectin-binding integrins $\alpha v\beta 3$ and $\alpha v\beta 5$. Modifications of the penton base have allowed other integrins to be targeted. For example, replacing the RGD motif with LDV directed the binding of recombinant adenoviruses to the integrin $\alpha 4\beta 1$.

It is evident that extremely complex manipulations are needed to achieve redirected viral targeting and until the fundamental information on the structure of these envelopes is available, the system remains 'hit or miss'.

8.4 Liposome vectors

An alternative to the viral delivery systems is the encapsulation of genetic material inside a liposomal carrier molecule. Liposomes have been used to deliver DNA into a variety of cell types, but the 'targeting' is generally via physical delivery, e.g. bronchial instillation for the treatment of cystic fibrosis or direct injection into melanomas. This is because liposomes are selectively taken up by cells of the reticuloendothelial system (RES), in particular, macrophages resident in the liver, spleen and bone marrow. This natural affinity for RES cells can however be exploited in some circumstances. For instance, in *L. donovani* leishmaniasis, parasites replicate in Kupffer cells of the liver and reside in a vacuole to which liposomes fuse. Liposomes in this instance are not only taken up passively by the target cell, but also target the appropriate organelle, making liposome-mediated delivery of transcriptionally targeted antisense or suicide genes a potential treatment. In most cases unless ligands or fusogenic moieties are coupled to liposomes (discussed below) it is diffcult to avoid uptake by the RES; although this can be delayed by the use of stealth liposomes displaying negatively charged groups such as ganglioside GM1 and polyethylene glycol. Similarly, some liposome formulations can be selectively taken up by certain tissues of organs. This presumably occurs because of differences in the protein/lipid composition of different cell membranes matching the incoming liposome. For instance, incorporation of a thiol-reactive lipid into a cationic lipid–DNA complex was able to improve targeting to hepatic cells 35-fold. Further modification of the thiol group to form a tetra-antennary galactolipid produced a complex that was capable of targeting hepatocytes through the asialoglycoprotein receptor, increasing the transfection efficiency in hepatic cells 1000-fold. It is however, widely recognized that specific targeting may be more readily achieved by the incorporation of targeting moieties into the liposome structure. Some of the pioneering work has involved the incorporation of monoclonal antibodies into the lipid bilayer, creating immunoliposomes which demonstrate novel tropisms conferred by the antibody or ligand. Examples include liposomes (containing the CAT reporter gene) coated with antibodies against the mouse histocompatability antigen H_2K^k. When these immunoliposomes were injected into immunodeficient mice bearing ascites tumors expressing this antigen, elevated levels of CAT expression were observed compared to delivery with normal liposome complexes. Similar results were seen using MAbs directed against glioma-associated antigens. Further studies have used antibodies to human placental alkaline phosphatase to target immunoliposomes to cultured epidermoid carcinoma cells expressing this enzyme and to Dalton's lymphoma as well as rat hepatoma cells.

In another approach peptide ligands such as transferrin have been used to deliver DNA in combination with liposomes, to erythroid cells in bone marrow, which express the transferrin receptor. Some carbohydrate moieties have also been used, such as β-galactosidase and lactose which can increase the targeting to hepatocytes. More recently the incorporation of integrin-targeting peptides into lipofectin/DNA complexes has improved targeting by over 50% in some cases. However, although these targeting strategies all look promising, specific fusion with the cell membrane is not enough to ensure delivery of the therapeutic gene to the nucleus and many of these liposomes are degraded by the endosomal pathway. To this end, components of enveloped viruses such as the F protein of Sendai virus (hemagglutinating virus of Japan [HVJ]) have been incorporated into liposomes to create 'virosomes' which couple the efficient entry characteristics of the virus with the safety of the liposome. Initial studies coupled liposomes encapsulting the HSVTK gene with HVJ; these particles were able to transduce 10% of the mouse L target cells *in vitro*. Subsequent improvements have included the incorporation of gangliosides in the liposome to serve as receptors for HVJ, together with a nuclear protein to compact the nucleic acid increasing both its encapsulation efficiency and ability to translocate to the nucleus. The HVJ method appears efficient, but will only be applicable where one dose of a therapeutic gene is required as the viral envelope genes on multiple injections will elicit an immune response.

In the future combinations of antibody targeting and use of viral fusion properties together with nuclear localization signals may lead to the development of targeted 'superliposomes' for gene delivery.

8.5 Molecular conjugates

One particular field which is generating particular interest is the development of molecular conjugates which deliver genes via the receptor-mediated endocytosis pathway. Conjugates use polycations, such as poly-L-lysine, to compact DNA into toroids. Targeting moieties such as antibodies can then be incorporated into these structures allowing them to interact with the negative charges on the DNA, leaving the ligand exposed on the surface of the conjugate. The ligands selected must be efficiently endocytosed to allow effective internalization of the DNA. One of the first receptors targeted was the asialoglycoprotein receptor expressed on hepatocytes which interacts with the asialoorosomucoid (ASOR). In the system described BSA was modified by galactosylation giving it specificity for the ASOR receptor. This has since been used to target CAT and human factor IX cDNAs to hepatoma cells and to liver *in vivo*. Similar strategies have been used with ligands such as lectins, EGF and transferrin. The major problem with this system is its reliance on the endosomal pathway for internalization, this results in the majority of the complexes being degraded in lysosomes resulting in poor transduction efficiency. To enhance delivery these conjugates have been covalently linked to replication defective adenoviruses, utilizing the viruses property for endosomal escape. Again a drawback with this system is the ubiquitous nature of the adenovirus receptor, abolishing the molecular conjugate's targeting potential. To overcome this adenoviruses are linked to the conjugates via the fiber protein, abolishing the normal viral targeting or by the addition of a monoclonal antibody to the fiber, neutralizing its effect. Such conjugates are more applicable to *ex vivo* use; first, because of size restrictions *in vivo* prohibiting extensive extravasation or tissue penetration, and second, the use of viral proteins makes the system immunogenic.

8.6 Targeting at the transcriptional level

Another targeting strategy exploits tissue or tumor specific promoter elements to drive the expression of a therapeutic gene in those cells that contain transcription factors able to activate the promoter, this in turn theoretically allows relatively promiscuous delivery of the vector. For this system to work the regulatory transcriptional elements need to be fully characterized and even under these circumstances it has become clear that correctly regulated expression can require elements distant to the promoter, some of which are located 3' of the coding region. These elements known as locus control regions (LCR) act in concert with the promoter to drive tissue specific expression independent of position of integration. As a consequence cis-acting regulatory sequences must be identified and condensed and arranged so that they fit within the vectors currently available. Many examples now exist where tissue specific promoters have been used to direct expression of therapeutic genes. One of the most well documented systems is for the treatment of cancer using genetic prodrug activation therapy (*Figure 8.3*), this utilizes tissue or

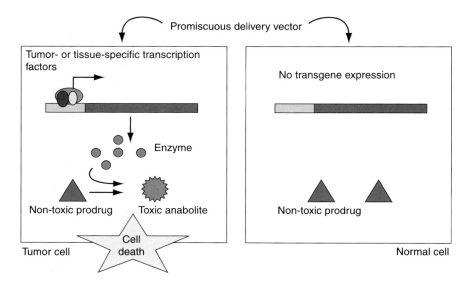

Figure 8.3. GPAT utilizing transcriptional targeting. Tissue/tumor specific promoters can only function in cells which have the relevant transcription factors to activate them. The suicide gene expressed in these cells sensitizes them to a nontoxic prodrug. Normal cells transduced with the promiscuous vector do not contain the relevant *trans*-activating factors therefore, cells remain immune.

tumor specific promoters (*Table 8.1*) to drive expression of a suicide gene only in those cells containing the correct array of transcription factors. The first system reported used the hepatoma associated α-fetoprotein (AFP) promoter to drive specific expression the varicella zoster virus-tk gene in hepatoma cells.

Table 8.1. Gene transcriptional elements used for GPAT systems

Gene	Tumor or tissue type
ERBB2	Breast, pancreatic and gastric tumors
ERBB3	Breast and gastrointestinal cancers
ERBB4	Breast and gastrointestinal cancers
MUC1	Breast, pancreatic duct tumors
Carcinoembryonic antigen (CEA)	Colorectal, pancreatic and gastric tumors
Bombesin	Small cell lung carcinoma
DOPA decarboxylase	Small cell lung carcinoma
Neuron-specific enolase	Small cell lung carcinoma
Tyrosinase	Melanoma
Tyrosinase-related protein (TRP-1)	Melanoma
Prostate-specific antigen	Prostate cancer
Secretory leukopeptidase inhibitor	Lung
Thyroglobin	Follicular carcinoma of the thyroid
Insulin	Insulinoma
11-β-hydroxylase	Adrenocortical carcinoma
α-Fetoprotein	Hepatocellular carcinoma

Human melanomas have provided the target for another tissue-specific suicide gene strategy. In this system the HSVTK gene was placed under control of the tyrosinase promoter (which is up-regulated in melanomas). Expression of the suicide gene was restricted to melanoma cell lines compared to control fibroblast. In *in vivo* experiments transfected murine melanoma cells expressing HSVTK were injected into immunocompetent syngeneic mice. On infusion with ganciclovir tumor regression was evident in 29 of the 30 animals. Although the system looks promising there are problems with the use of the tyrosinase promoter, as other cell types such as spinal ganglia, astrocytes and Schwann cells also express tyrosinase, which may result in normal cells being sensitized to ganciclovir. A further study has exploited promoter elements of the proto-oncogene ERBB2. This is over-expressed in a number of cancers such as pancreatic and breast carcinomas. The transcriptional subunits implicated in its up-regulation in breast cancer have been localized to a site lying 500 bp proximal within the promoter which binds the transcription factor AP2. This 500 bp fragment of the promoter was used in a retroviral construct to drive expression of the cytosine deaminase gene. A panel of cells with varying ERBB2 status were transduced and treated with 5FC. The proportion of cell death was directly related to the ERBB2 expression level of the cells.

In recent years there have been several reports of the development of chimeric promoters in an attempt to improve transcriptional targeting. One study has exploited the use of the Myc–Max oncogene found to be up-regulated in a number of cancers. The protein heterodimer is able to bind the motif CACGTG which switches on transcriptional activity. A construct was prepared which contained four repeats of this motif proximal to a minimal promoter driving a luciferase reporter gene. Myc-positive cells transfected with the construct showed a 4–6-fold increase in expression of the luciferase gene compared to transfected Myc-negative cells. The reporter gene was replaced with a suicide gene encoding the enterotoxin gene. Established Myc-positive tumors in nude mice were injected with the suicide construct, or a control construct expressing the β-galactosidase gene. There was a 50% decrease in tumor burden in the *Pseudomonas* toxin-treated animals compared to the controls. Subsequent studies replacing the toxin gene with the HSVTK gene have proved similarly successful.

One of the first examples describing the rearrangement of transcriptional elements to produce an 'enhanced' tissue specific promoter has utilized the carcinoembryonic antigen gene (CEA). CEA is a tumor associated cell surface antigen that is expressed in normal tissue but upregulated in tumor tissue such as lung and colorectal cancer. By combining different CEA promoter/enhancer sequences it was possible to increase tissue specific expression of a reporter gene 2–4-fold over normal expression from a constitutive SV40 promoter. It would appear that the development of chimeric promoters is the key to transcriptional targeting.

One thing that is very clear from these studies is that relevant control regions lie over many kilobases of sequence and within chromatin domains that are often difficult to copy in the current vector systems. In addition, transcriptional combinations may be successful in some vectors and not others. For instance viral delivery systems are not always efficient and transgene expression can be lost or compromised. This may be attributed to: (1) partial loss of tissue specificity from the internal promoter, (2) promoter interference, and (3) reduced virus titers due to the large minigenes inserted.

Similarly, reports documenting loss of tissue specific expression from adenoviral vectors have been noted. It is envisaged that in this system strong cryptic promoters may override weak tissue specific transcriptional elements. One way to alleviate this is to incorporate insu-lating sequences around the transgene cassette 'protecting' from intervening adenoviral sequences.

It is clear that careful consideration of the *cis* acting control elements and the vectors used need to be addressed. New emerging vectors such as the 'gutless' adenovirus (which has the capacity to package over 30 kb of sequence) may alleviate some of the size restrictions, allowing packaging of much larger transcriptional control elements.

8.6.1 Transgene function – a further level of targeting

A further level of targeting may be achieved by selecting suitable therapeutic genes. For instance in the case of GPAT it is impossible to target every cell with a suicide gene. Hence the system is reliant on a strong 'bystander effect'. This phenomenon has been noted in mixed populations of cells (some transduced with a suicide gene construct, while others are not). When treated with the respective prodrug death results in both transduced and nontransduced cells. This feature has proved to be extremely advantageous where the transduction efficiency is poor. 'Bystander' effect results from the trafficking of small molecules (molecular weight < 1000) between cells via gap junctions, phagocytosis of apoptotic toxin filled vesicles and, more importantly, a T-cell mediated immune response, effectively amplifying the killing mechanism. The major challenge for all cancer gene therapists is how to treat distant metastases. The most obvious route is to stimulate the host's immune response to recognize the 'danger'. This is being addressed by in-corporating immunostimulatory genes into suicide cassettes and attempting to shift the targeting and therapeutic amplification to the host's immune response.

8.7 Conclusion

It is clear that none of the targeting strategies alone provide the ultimate gene therapy vector, but by combining facets from each of this strategies it may be possible to produce tailor-made targeted vectors suitable for each disease. Similarly with the intensive research invested in understanding viral life cycles, protein engineering, crystallography and molecular pathology it is only a matter of time before the ultimate gene delivery vehicle is created.

Further reading

Cosset, F.L. and Russell, S.J. (1996) Targeting retrovirus entry. *Gene Ther.*, **3**, 946–956.

Gerrard, A.J., Hudson, D.L., Brownlee, G.G. and Watt, F.M. (1993) Towards gene therapy for haemophilia B using primary keratinocytes. *Nat. Genet.*, **3**, 180–183.

Martin, L.-A. and Lemoine, N.R. (1996) Direct cell killing by suicide genes. *Cancer Metastasis Rev.*, **15**, 301–316.

Miller, N. and Whelan, J. (1997) Progress in transcriptionally targeted regulatable vectors for genetic therapy. *Hum. Gene Ther.*, **8**, 803–815.

Russell, S.J. (1994) Replicating vectors for gene therapy of cancer: risks, limitations and prospects. *Eur. J. Can.*, **30A**, 1165–1171.

Van Beusechem, V.W. and Valerio, D. (1996) Gene transfer into hematopoietic stem cells of nonhuman primates. *Human Gene Ther.*, **7**, 1649–1668.

Vile, R.G. and Chong, H. (1996) Immunotherapy: combinatorial molecular immunotherapy—a synthesis and suggestions. *Cancer Metastasis Rev.*, **15**, 351–364.

Chapter 9

Gene therapy in the clinic: human trials of gene therapy

Thomas Valere

9.1 Introduction

Broadly speaking, in the 1970s, ever-increasing attention was paid to the biochemical alterations underlying human disorders such as hereditary diseases, cancer and degenerative disorders. The origin of more and more diseases was traced to the molecular level.

In the 1980s a number of molecular tools for detecting genetic alterations have become available and widely used, such as gene mapping, sequence determination, the Southern blot technique, and the polymerase chain reaction (PCR). Concurrently, substantial advances were being made in virology, which enabled new and powerful tools to be developed for transferring genes.

9.2 Gene therapy around the globe

Clinical gene therapy is in its infancy. In the late 1980s a new form of human therapy, based on the administration of genetic material, had become a real possibility. The first protocol for human somatic transgenesis was approved in the USA by the Recombinant DNA Advisory Committee (RAC) on 3 October 1988 and by the National Institutes of Health (NIH) on 2 February 1989. The trial was initiated a few months later, on 22 May 1989 [1]. The following decade has seen an explosion of gene therapy techniques and clinical protocols.

The figures in this chapter illustrate a world overview today – who does what in which country. North America remains the leader by far, with the USA accounting for at least 70% of the world's trials and patients (*Figure 9.1*). Trials are also under way in Canada. Europe, including Eastern European countries such as Poland, has been initiating

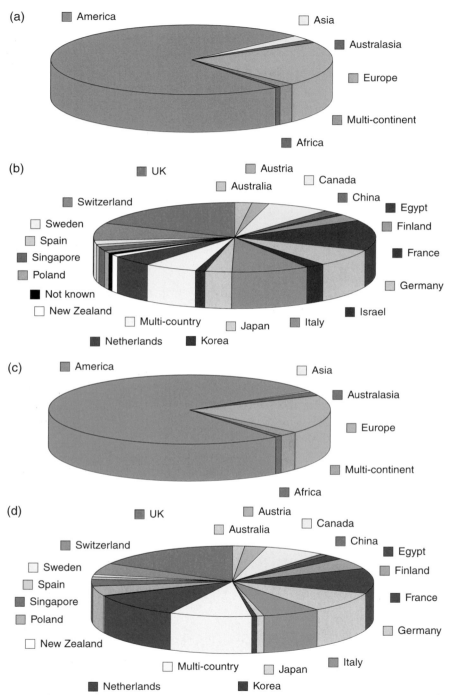

Figure 9.1. (a) Protocols by continent. (b) Protocols by country outside the USA. (c) Patients by continent. (d) Patients by country outside the USA. Total number of protocols = 351. Total number of patients = 2687.

trials at an increasing rate since 1995. China was the first Asian country to begin gene therapy in 1991, followed in 1995 by Japan, and in April 1997 by Korea. Two trials have also been started in Israel and two in Singapore.

9.3 Vectors and routes of administration

Gene transfer problems call for many solutions. In man, the vectors and routes used are chosen according to the clinical goal pursued, and according to available preclinical data.

Historically, retroviral vectors were the first chosen. They were used in all the trials until 1991. Although new trials are increasingly promoting adenovectors and liposome/polycations, retroviral vectors remain the principal gene therapy tools.

Until 1996, the vast majority of gene transfers were done *in vitro*. In 1997 an increasing number of *in vivo* administrations was seen. Today, almost half the patients treated undergo *in vivo* gene transfer. The pie charts in *Figure 9.2* give a breakdown of the numbers of patients and protocols by vector and route of administration.

9.4 Which genes for which diseases?

The number of genes and gene combinations transferred into humans is approaching 100. These are categorized in *Figure 9.3*. The full list of all these genes, with updated numbers of protocols and patients are permanently available on the internet at http://www.wiley.com/genmed/clinical/genes.html

Monogenic diseases were the first targets considered for gene therapy. Since 1990, the majority of requests for trial approval have been in the area of oncology, involving both gene marking studies and true gene therapy studies. A number of the latter are in progress and some results have been published. The studies may be classified into several categories.

(i)   Transfer of suicide genes, which confer drug sensitivity to cancer cells. Transduced cells become highly sensitive to the corresponding drug and can be selectively eliminated.
(ii)   Transfer of antisense genes to block the expression of deleterious cancer-promoting genes.
(iii) Transfer of drug resistance genes into nonmalignant blood stem cells to protect them from subsequent chemotherapy. Once physiological stem cells are protected, more aggressive chemotherapy can be employed to eliminate tumor cells.
(iv)   Transfer of tumor suppressor genes into cancer cells to replace a missing or damaged cancer-blocking gene.
(v)   Transfer of recombinant antibody genes to interfere with tumor cell specific functions.

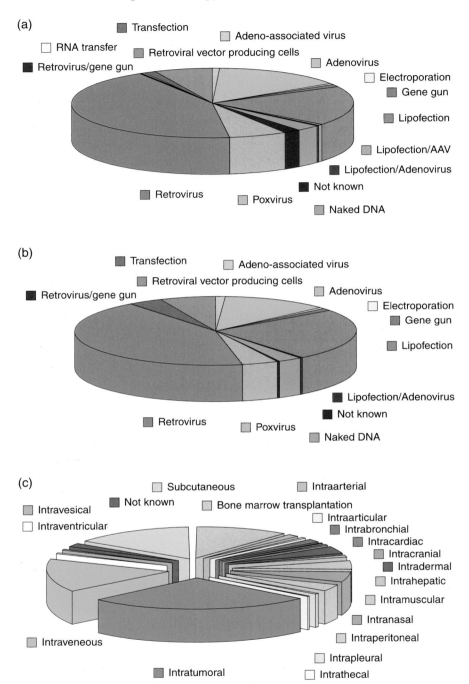

Figure 9.2. (a) Protocols by vector. (b) Patients by vector. (c) Protocols by route. (d) Patients by route. Total number of protocols = 351. Total number of patients = 2687.

(d)

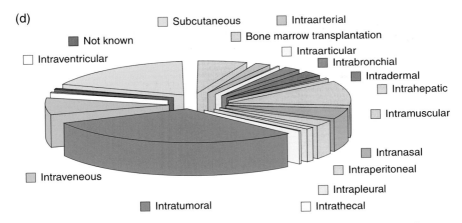

Subcutaneous • Intraarterial • Not known • Bone marrow transplantation • Intraventricular • Intraarticular • Intrabronchial • Intradermal • Intrahepatic • Intramuscular • Intranasal • Intraperitoneal • Intraveneous • Intrapleural • Intratumoral • Intrathecal

Figure 9.2. Continued.

(vi) Transfer of genes coding for proteins, which boost the host's antitumor immune response. Nabel's protocol was the first to show that transferring the expression of allogeneic MHC proteins to *in situ* tumors increased tumor rejection. Additionally, CTLs specific to the untransduced autologous tumor may be generated by this immunotherapy.

(vii) Transfer of oncogene down-regulating genes to shut off cancer-inducing genes.

9.5 Results: the global picture

First, the vast majority of investigators reported the absence of serious side-effects due to gene transfer. Second, feasibility of human gene therapy is now clearly established. On the other hand, the majority of teams detected actual gene transfer, but rarely a durable one, and even more rarely with a durable therapeutic benefit. Clinical improvements, if observed, are generally transient. But some promising durable treatments have still been achieved in specific cases.

A summary of all these results is given below. The full text, regularly updated with new results, and/or bibliographical references, is available at http://www.wiley.com/genmed/clinical. A breakdown of the total number of protocols and patients by disease is given in *Figure 9.4*.

9.5.1 Gene marking studies

It was initially unknown whether human somatic transgenesis would be safe, whether the transduced genes would be durably expressed, and whether the transduction process would affect natural cell properties.

For example, would tumor-infiltrating lymphocytes (TIL), which exhibit natural *in vivo* tumor tropism, continue to do so when transduced?

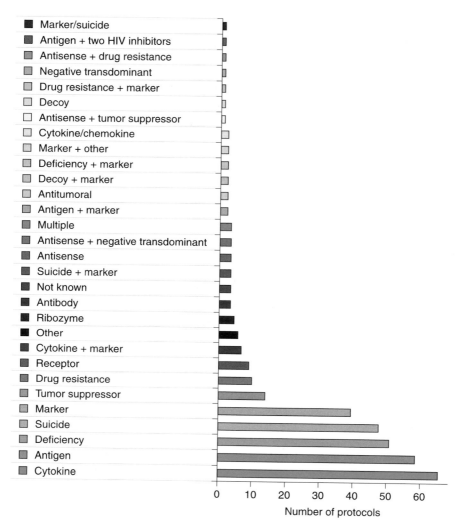

Figure 9.3. Protocols by gene type. Total number of protocols = 351. Total number of patients = 2687.

If transduced cells conserved their initial properties, interesting applications could be envisaged. Rosenberg *et al.* put this to the test when they used transduced TIL for tumor-targeted anticancer drug-delivery purposes [1]. This gene marking clinical trial was the first attempt at human somatic transgenesis. It was approved by the RAC on 3 October 1988 and by the NIH on 2 February 1989. The trial was initiated a few months later, on 22 May 1989. Notably, it is the only trial with no closing date nor any legal limitation on the number of patients. Nevertheless, it enrolled only 10 patients, and the results were sufficiently informative to establish the feasibility of human gene therapy.

(a)

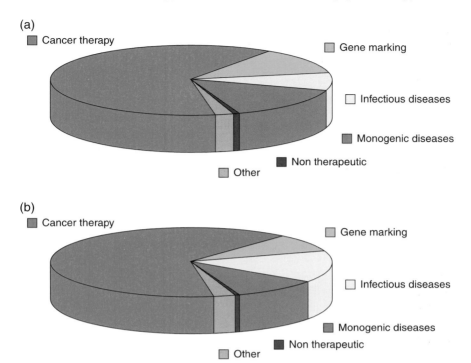

(b)

Figure 9.4. (a) Protocols by disease. (b) Patients by disease. Total number of protocols = 351. Total number of patients = 2687.

Five patients with malignant melanoma received 1–3 transfusions of around 100 billion autologous TILs, previously marked *ex vivo* with a retroviral vector expressing NeoR, a reporter gene. No helper viruses were found, and no reverse transcriptase activity was detected in TIL samples. No toxicity of treatment was observed, and the TIL remained IL-2 dependent. The transgene was detected by PCR in peripheral blood mononuclear cells 3 weeks after transfusion in all patients, and up to 2 months after transfusion in two patients. The transgene was also detected in tumor biopsies until day 64. This demonstrated that *ex vivo* transduction does not abolish the natural tumor-targeting properties of the TIL.

Using similar technology, Brenner *et al.* observed that blood progenitor cells could be transduced too, demonstrating that autologous marrow infusion restores long-term hemopoiesis in cancer patients, and that blood cells can be modified by targeting their progenitors [2].

Dunbar *et al.* marked and traced two CD34-enriched populations of cells: peripheral blood cells and bone marrow cells [3]. The marker gene was detectable in all 10 patients able to be evaluated at transplantation time, and in three out of nine patients 18 months later. Their informative protocol demonstrated that peripheral blood cell grafts appeared to result in better long-term marking than bone marrow cells.

Safety of autograft procedures. Tumor relapse is the most frequent cause of death following autologous bone marrow transplantation. This may be because some tumor cells are not eradicated by therapy, but there is also a case for arguing that harvested autologous bone marrow may contain tumorigenic cells. These cancer cells would be amplified during the *ex vivo* phase, before reinjection into the patient.

Rill *et al.* used gene-marked grafts in eight neuroblastoma patients [4]. Their observations suggest that grafts do contain, and should be purged of, malignant cells prior to reinjection, possibly in all cases of marrow-infiltrating solid tumors.

Assessment of bone marrow purging efficiency. The method Rill *et al.* described allows control and amelioration of purge techniques, through marker tracing in small-scale clinical trials [4]. Such trials are currently under way.

9.5.2 Gene therapy of cancer

Immunotherapy. A gene coding for the HLA-B7 glycoprotein was transferred into metastatic melanoma tumors by direct injection of a DNA-liposome complex in five patients (all HLA-B7 negative). Tumor biopsies revealed HLA-B7 protein and its corresponding transferred plasmid DNA. In all five patients, antitumor (autologous) and anti-HLA-B7 (allogeneic) cytotoxic lymphocyte immune responses were detected. In one patient, two injected nodules regressed completely, as well as other distant noninjected metastatic lesions.

It is interesting to observe that this protocol, proposed by Nabel *et al.* in 1992, is now implemented in about 30 trials worldwide, with more than 300 patients [5]. These are in predominantly phase II (70%) or phase I/II (24%) trials. Few results were yet reported; Rubin *et al.* reported that tumor regressions were observed in six out of 15 patients treated [6].

Stopeck *et al.* noted that injected nodules regressed by 25% or more (radiologic and physical examination) in seven patients out of 17 [7]. One patient with a single nodule achieved complete remission. In their trial, toxicities such as pain, hemorrhage, pneumothorax and hypotension were observed. The authors attribute these effects to technical aspects of the injections or biopsies.

Suicide genes. Although numerous teams perform such trials, few results are yet public. Weber *et al.* conducted an open-label, multicenter phase II trial in Germany and reported a 1-year follow-up of 10 patients with recurrent glioblastoma multiforme [8]. Retroviral vector packaging cells were administered either by stereotactic injection into tumors, or during open tumor surgery. The murine cell line produced retroviral vectors coding for the TK suicide gene. Fourteen days after tumor debulking and cell administration of one billion cells in 50 single injections, ganciclovir (GCV) treatment was started for 14 days. No adverse effects related to intracerebral administration of murine cells were seen and treatment was well tolerated. Four patients died of tumor progression. Of the other six patients, one presented a complete remission at 12 months, and five had progressive disease, though with a significant quality of life. Tumor regrowth was detected in areas which may have been insufficiently resected or insufficiently cell-injected.

Izquierdo *et al.* reported that patient survival time increased considerably if neurosurgical tumor size reduction was performed in addition to suicide gene therapy [9].

Bordignon *et al.* used the suicide option to control graft-versus-host disease (GVHD) [10]. Allogenic bone marrow transplantation is a useful procedure for immunotherapy of hematological malignancies in general, and of some nonmalignant blood disorders. However, in cases where high T-cell doses need to be administered, a severe problem occurs in about half the patients. Allogenic lymphocytes can mediate GVHD with life-threatening consequences. In order to create the option of eliminating donor cells *in vivo* should GVHD occur, Bordignon *et al.* attempted to transduce close to 100% of the donor cells with a suicide gene, TK, before transplantation [10]. GVHD was detected in three out of eight patients who were able to be evaluated. They were treated with GCV injections. In one patient, the donor cells could never be totally eradicated despite extended GCV treatment. GVHD was attenuated but not eliminated. In the other two patients, GCV administration rapidly destroyed all donor cells. A sharp 24-hour decrease from 13% of allogenic lymphocytes to PCR-undetectable levels was observed, in the absence of any immunosuppressive drugs, and without toxicity. This totally abolished the immune attack, and all clinical and biological signs of GVHD disappeared in both cases.

The authors emphasize that no such remissions from this stage of GVHD have ever occurred spontaneously among 258 patients examined in an international study of classical (nongene therapy) allogenic bone marrow transplantation.

9.5.3 Gene therapy of monogenic diseases

Adenosine deaminase deficiency (ADA–SCID). Historically, this was the first gene therapy clinical trial which actually aimed to cure a disease. On 14 September 1990, Culver *et al.* initiated a clinical trial in ADA-deficient children for whom alternative therapies (PEG-ADA administration) had failed to restore an adequate immune system [11]. This clinical trial is now in phase I/II with several target cell options (lymphocytes, CD34+ cells, blood stem cells of newborns).

Peripheral blood lymphocytes from two children were collected and expanded *ex vivo*. They were infected with a retroviral vector expressing the human ADA. Seven hundred million autologous lymphocytes per kilogram of body weight were infused. The same operation was repeated monthly, and the last infusions took place in October 1992.

Since then, both children have demonstrated ADA + circulating lymphocytes, as shown by PCR, Southern blotting and protein immunodetection. The previous alternative treatment, PEG-ADA administration, was continued but the dose was reduced by 50% for both children. One had over 50% ADA + lymphocytes post-treatment and her immune state was normalized. The other girl's lymphocytes turned out to be 10 times less susceptible to infection with the vector. Post-treatment, her blood showed only 0.1–1% ADA + lymphocytes, but her clinical state remained satisfactory.

In April 1993, several gene therapy trials in newborns were approved in the USA. Autologous placental and umbilical cord cells were transduced *ex vivo* by the ADA retroviral vector. Unlike peripheral blood, placental and umbilical cord blood is naturally enriched in blood stem cells. This raised hope that a single intervention would suffice to treat ADA–SCID newborns definitively.

Two years post-treatment, authors Kohn *et al.* reported that different cell lineages exhibited the ADA integrated sequences [12]. The neo gene, also present in the retroviral vector, was expressed as well. Since sufficient ADA expression was observed, PEG-ADA administration was progressively reduced under regular biomedical supervision. The team led by Bordignon *et al.* reported similar results [13].

The first Japanese gene therapy trial was very similar to the initial American ADA trial and used the same vector. After extensive official evaluation in Japan, the trial was approved on 13 February 1995. The patient, a 3-year-old boy with ADA–SCID, was treated on 1 August 1995. Onodera *et al.* report a 1-year follow-up [14]. During these 12 months, ADA + peripheral blood lymphocytes (PBL) remained stable at 10–20% of total PBLs. Prior to treatment, only marginal ADA activity was observed in the patient's PBLs. Post-treatment, the ADA activity levels were comparable to those of a heterozygous ADA + /ADA − carrier. Cell counts showed increased T-cell numbers, and the patient's immune functions were improved.

Hoogerbrugge *et al.*, describing a related, but not identical, multi-country trial reported that ADA gene presence was detected for up to 3 months in the blood and for up to 6 months in the marrow, but that no gene expression was detected [15].

Familial hypercholesterolemia. This condition was addressed by Grossmann *et al.* in a one-patient trial [16]. Hypercholesterolemia was partially corrected by hepatocyte autograft after *ex vivo* transfer of the correcting gene (LDL receptor gene). Expression was stable for the 18-month follow-up period. The LDL/HDL ratio dropped from 10 pretreatment to 5 post-treatment. Clinically, the authors report that coronary disease did not progress during the 18 months of follow-up. No conclusion about the long-term benefit of treatment can be drawn. The authors proposed to continue clinical trials using this protocol, and possibly to extend the trials later to genetic metabolic liver disorders.

Cystic fibrosis. The feasibility of adenovirus-mediated gene therapy was first tested on 17 April 1993, in a clinical trial conducted by Crystal *et al.* involving four patients with cystic fibrosis (defective CFTR function) [17]. Two billion adenovirus-CFTR plaque forming units (pfu) were administered to the nasal or bronchial epithelium. For 7 days following administration, CFTR protein and mRNA, undetectable prior to treatment, were observed in about 14% of one patient's epithelial cells, and mRNA alone in another patient. After 10 days expression was no longer detectable. No adverse effects due to the adenovirus were observed during the 6–12-month follow-up period.

More than 20 teams, in several countries, have attempted cystic fibrosis correction using the CFTR gene in an adenoviral or liposomal/polycationic vector. Although 80% of these trials were initiated before December 1995, they have remained in phase I (14) or I/II (10) trials, but none has reached phase II. Indeed, the published phase I results were not as encouraging and decisive as hoped. Gene transfer and expression were observed, but always transiently, for a couple of weeks at most. Some partial electrophysiological corrections were noted in several trials, but they were transient too, and no significant clinical improvement, even transient, has yet been reported.

9.5.4 Gene therapy of infectious diseases (AIDS)

It is interesting to note that, as a rule, AIDS gene therapy trials recruit more than twice as many patients as non-AIDS gene therapy trials. In particular, the only gene therapy trials which have recruited over 100 patients (and even over 200) are the AIDS trials. Nevertheless, few results have been published, and they did not mention patient improvement.

The open-label phase I/II trial of Haubrich *et al.* was conducted on seropositive adults [18]. Either a placebo or a particular retroviral vector, HIV-IT(V), was injected intramuscularly into patients. HIV-IT(V) is highly recombinant and is produced in canine cell lines, as opposed to usual murine packaging lines, in order to lower *in vivo* recombination risks. It carries segments of the potentially vaccinating *env* and *rev* genes of HIV-1 IIIB. Observation of the first 16 patients revealed no toxicity. No therapeutic effect was reported, in terms either of vaccination or of reduction of the viral load.

This trial and its phase I predecessor are now closed. They have been followed by a related open-label phase II trial which also uses HIV-IT(V). The trial recruited 124 patients in the first months of 1995 and recruitment reached 216 patients in 197.

McGregor *et al.* reported a randomized trial, conducted by Weiner [19]. It was the first instance of naked HIV genes injected into patients. Intramuscular injections of vaccine candidate DNA expressing HIV-1 MN *env* and *rev* genes were given to 13 men and two non-pregnant women, but no statistically significant effect of the treatment could be shown.

Riddell *et al.*, in another trial using TK, observed a rapid rejection of trans-gene-expressing autologous cells, and no therapeutic benefit was seen [20].

9.5.5 Other indications

Amyotrophic lateral sclerosis was addressed by using baby hamster kidney cells (BHK cell line) transfected *in vitro* with the ciliary neurotrophic factor (CNTF) gene and encapsulated in polymer. Capsules were then implanted in the lumbar intrathecal space of patients. Although CNTF expression was detected *in vivo*, the investigators observed that the disease continued to progress in all patients.

9.6 Conclusion

Some promising clinical observations in certain phase I trials encouraged recruitment for phase II evaluation. One example is the phase I cancer trial performed by Gary Nabel [6] which has shown interesting clinical results on cancer.

The original protocol has since been resubmitted for approval in a similar or slightly modified version, by the same and other investigators. Given the momentum created by pre-existing results, these protocols were approved more readily by the regulatory authorities.

This 'follower effect' can also be noted for the encouraging ADA–SCID results, though to a lesser extent since there is a much greater need for cancer treatments than for any monogenic disease.

The same phenomenon might also occur with cancer trials using HSV-TK cell therapies, which are frequently duplicated now. Most of these trials are now in phase I/II. It is noteworthy that HSV-TK cell therapy is the first approach to have reached phase III.

Acknowledgments

This chapter is adapted from *The Journal of Gene Medicine* website located at http://www.wiley.com/genmed, © 1998–9, John Wiley & Sons.

References

1. Rosenberg, S., Aebersold, P., Cornetta, K. *et al.* (1990) Gene transfer into humans – immunotherapy of patients with advanced melanoma, using TIL modified by retroviral gene transduction. *N. Engl. J. Med.*, **323**, 570–578.

2. Brenner, M., Rill, D., Holladay, M. *et al.* (1993) Gene marking to determine whether autologous marrow infusion restores long-term haemopoiesis in cancer patients. *Lancet*, **342**, 1134–1137.
3. Dunbar, C.E., Cottler-Fox, M., O'Shaughnessy, J.A. *et al.* (1995) Retrovirally marked CD34-enriched peripheral blood and bone marrow cells contribute to long-term engraftment after autologous transplantation. *Blood*, **85**, 3048–3057.
4. Rill, D., Santana, V., Roberts, W. *et al.* (1994) Direct demonstration that autologous bone marrow transplantation for solid tumors can return a multiplicity of tumorigenic cells. *Blood*, **84**, 380–383.
5. Nabel, G.J., Nabel, E.G., Yang, Z.Y., Fox, B.A., Plautz, G.E., Gao, X., Huang, L., Shu, S., Gordon, D. (1993) Direct gene transfer with DNA-liposome complexes in melanoma: expression, biologic activity, and lack of toxicity. *Proc. Natl Acad. Sci. USA*, **90**, 11307–11311.
6. Rubin, J., Charboneau, J.W., Reading, C. and Kovach, J.S. (1994) Phase I study of immunotherapy of hepatic metastases of colorectal carcinoma by direct gene transfer. *Hum. Gene. Ther.*, **5**, 1385–1399.
7. Stopeck, A.T., Hersh, E.M., Apkoriaye, E.T., Harris, D.T., Grogan, T., Unger, E., Warneke, J., Schluter, S.F. and Stahl, S. (1997) Phase I study of direct gene transfer of an allogeneic histocompatibility antigen, HLA-B7, in patients with metastatic melanoma. *J. Clin. Oncol.*, **15**, 341–349.
8. Weber, F., Bojar, H., Priesack, H.B., Floeth, F., Lenartz, D., Kiwit. J. and Bock, W. (1997) Gene therapy of glioblastoma – one year clinical experience with ten patients. *J. Mol. Med.*, **75**, B40 (126).
9. Izquierdo, M., Cortes, M.L., Martin, V., de Felipe, P., Izquierdo, J.M., Perez-Higueras, A., Paz, J.F., Isla, A. and Blazquez, M.G. (1997) Gene therapy in brain tumours: implications of the size of glioblastoma on its curability. *Acta Neurochir. Suppl.* (Wien), **68**, 111–117.
10. Bordignon, C., Bonini, C., Verzeletti, S. *et al.* (1995) Transfer of the HSV-tk gene into donor peripheral blood lymphocytes for *in vivo* modulation of donor anti-tumor immunity after allogeneic bone marrow transplantation. *Hum. Gene. Ther.*, **6**, 813–819.
11. Culver, K., Anderson, F. and Blaese, R. (1991) Lymphocyte gene therapy. *Hum. Gene. Ther.*, **2**, 107–109.
12. Kohn, D., Weinberg, K.I., Parkman. P. *et al.* (1993) Gene therapy for neonates with ADA-deficient SCID by retroviral-mediated transfer of the human ADA cDNA into umbilical cord CD34+ cells. *Blood*, **82** (suppl. 1), 315a (Abstract 1245).
13. Bordignon, C., Notarangelo, L.D., Nobili. N. *et al.* (1995) Gene therapy in peripheral blood lymphocytes and bone marrow for ADA-immunodeficient patients. *Science*, **270**, 470–475.
14. Onodera, M., Ariga, T., Kawamura, N. *et al.* (1998) Successful peripheral T-lymphocyte-directed gene transfer for a patient with severe combined immune deficiency caused by adenosine deaminase deficiency. *Blood*, **91**, 30–36.
15. Hoogerbrugge, P.M., Valerio, D., Levinsky, R.J. *et al.* (1996) Bone marrow gene transfer in three patients with adenosine deaminase deficiency. *Gene Ther.*, **3**, 179–83.
16. Grossmann, M., Raper, S., Kozarsky, K., Stein, E., Engelhardt, J., Muller, D., Lupien, P. and Wilson, J. (1994) Successful *ex vivo* gene therapy directed to liver in a patient with familial hypercholesterolemia. *Nat. Genet.*, **6**, 335–341.
17. Crystal, R., McElvaney, M., Rosenfeld, M. *et al.* (1994) Administration of an Adenovirus containing the human CFTR cDNA to the respiratory tract of individuals with cystic fibrosis. *Nat. Genet.*, **8**, 42–50.

18. Haubrich, R. and McCutchan, J.A. (1995) An open-label, phase I/II clinical trial to evaluate the safety and biological activity of HIV-IT (V) (HIV-1 IIIB env/rev retroviral vector) in HIV-1-infected subjects. *Hum. Gene Ther.*, **6**, 941–955.
19. MacGregor, R.R., Boyer, J.D., Ugen, K.E. *et al.* (1998) First human trial of a DNA-based vaccine for treatment of human immunodeficiency virus type 1 infection: safety and host response. *J. Infect. Dis.*, **178**, 92–100.
20. Riddell, S.R., Elliott, M., Lewinsohn, D.A. *et al.* (1996) T-cell mediated rejection of gene-modified HIV-specific cytotoxic T lymphocytes in HIV-infected patients. *Nat. Med.*, **2**, 216–223.

Further reading

General public reviews, books and original articles cited in this chapter are referenced on *The Journal of Gene Medicine* website located at http://www.wiley.com/genmed, © 1998–9, John Wiley & Sons.

Chapter 10

Ethical issues in gene therapy

Norman C. Nevin

10.1 Introduction

Gene therapy has the potential to revolutionize clinical medicine. As originally conceived, gene therapy was an approach of transferring a normal copy of a single defective gene so that the disease would revert to a normal phenotype. Initially, it was seen as an approach to the treatment of single gene disorders. However, as gene therapy has evolved, it became clear that gene transfer could also be an approach for the treatment of both acquired diseases, such as cancer, and infectious diseases, such as AIDS. Gene therapy is broadly defined as the transfer of genetic material, DNA, to cells of an individual for therapeutic purposes. Although gene therapy has a long way to go before becoming conventional routine clinical management, there is no doubt, of its eventual success. Scientific advance, adherence to high standards of clinical research and ethical principles will lead to the success of gene therapy.

In considering the ethical issues, the initial question raised, was in what way does the use of human gene therapy differ from other conventional therapeutic interventions? [1] Some have assumed that the methods of human gene therapy are different, both in its nature and possible consequences, from other approaches to the management of disease. However, others consider human gene therapy as a natural extension of current techniques of clinical therapeutic management. The Clothier Committee whilst acknowledging that the primary intention was to benefit an individual patient, concluded that gene therapy should be regarded as research involving human subjects and should be governed by requirements similar to those which govern this kind of research [1]. Any research protocol to undertake human gene therapy should be subject to scientific, medical and ethical appraisal before its introduction. The ethical conduct of medical research is, in its broadest sense, the shared responsibility of all. Research on human subjects should conform to accepted ethical codes, the purposes of which are to maintain ethical standards in practice, protect subjects of research from harm, and preserve the subjects' 'rights' and liberties. Reassurance to the public and to professions that these are being done should be provided [2].

10.2 Supervision of gene therapy research

In the United Kingdom, the Committee on the Ethics of Gene Therapy made recommendations on the supervision of gene therapy [1]. Gene therapy may involve the deliberate modification of the genetic material of either somatic cells or germline cells. Germline gene therapy involves the introduction of a normal gene into germ cells (ova, sperm or the cells from which they are derived), such that it will correct the genetic defect and be transmitted in a Mendelian fashion from generation to generation. Somatic cell gene therapy is the insertion of new genetic material into the patient's cells, particularly those with primary disease pathology. It has been recommended that germline gene therapy should be prohibited [1], because little is known about the possible consequences and risks – particularly the harm to future generations which could result from such treatment. Approval was given for somatic cell therapy, as treatment would involve the affected individual and not future generations.

In order to regulate gene therapy, a supervisory committee was established with the necessary collective expertise to be able to assess the scientific merits and the ethical considerations of proposed research protocols. Such assessments should be made in conjunction with local research ethical committees. The supervisory body, Gene Therapy Advisory Committee (GTAC), was established in November 1993. The responsibilities of GTAC are listed in *Table 10.1*.

Gene therapy research proposals must be subject to ethical review by GTAC and local research ethical committees. There is a need for careful assessment of the scientific merits of the proposals, the competence of those undertaking the research, the possible risks and benefits of the

Table 10.1. Responsibilities of GTAC

(a)	Receive proposals from doctors who wish to conduct gene therapy in patients
(b)	Advise on content of the proposals, the design and conduct of the research
(c)	Advise on the facilities and service arrangements necessary for the conduct of the research
(d)	Act in co-ordination with local research ethical committees in advising on the ethical issues raised by the proposal
(e)	Establish and maintain a confidential register of patients who have been subjects of gene therapy trials
(f)	Act as a repository of up-to-date information on gene therapy internationally
(g)	Advise Health Ministers in the United Kingdom on scientific and medical developments which bear on the safety and efficacy of human gene therapy

proposed gene transfer therapy. Particular attention is given to informed consent ensuring that the patient understands the nature of the proposal, and the risks, benefits and obligations. Careful consideration is given not only to the content of the information provided to the patient, but also to the manner in which the information is given. This is necessary in such a complex field as gene therapy. The patient's interests and health must be paramount. When the physician suspects that the interest of the patient is at risk, the patient must be advised accordingly. The patient must be able to provide informed consent.

Each gene therapy proposal undergoes a rigorous process of assessment. Following internal assessment of the research protocol by GTAC secretariat, external reviewers, who are experienced in specific aspects of the research proposal, assess each application. On completion of external review, the researchers are notified of the issues raised in the external reviewers' reports and given the opportunity to modify their proposal accordingly. The researchers are invited to make an oral presentation to GTAC and to respond to the issues raised by their research protocol. The committee can give approval to the research proposal, subject to a satisfactory response to the questions raised. A proposal may be deferred until specific issues are resolved and/or further experimental work is undertaken. Proposals may be rejected, usually because they are unethical or the science is unsound [3]. A condition of approval is that GTAC must receive regular updates on the progress of the research and on modifications to the original proposal.

10.3 Consent to research

An important ethical requirement for gene therapy research involving patients is the consent of the patient. Patients should be invited to participate in research as volunteers. They are entitled to give or withhold consent to participate in any research project. Consent implies that sufficient information has been provided for the patient in a way that they can understand and enabled to make a voluntary decision, whether or not to participate in the proposed research. As gene therapy has novel, complex and far-reaching aspects, there is a concern to ensure that the patient is enabled to take these fully into account when giving consent. It is important to pay attention not only to the content of the information given to the patient but also to the way in which the information is conveyed. The risks, benefits and obligations must be carefully explained and the patient provided with an opportunity to ask questions. It is helpful if an independent source of advice can be obtained to ensure the patient understands the implications of giving consent. This person should not be a member of the research team and should have a sound grasp of the implications of gene therapy. It must be made clear to the patient that he/she can decline to participate without giving a reason and that the decision to decline will not disadvantage the patient's future clinical management.

There will be situations, by the nature of the disorder, or progression of the disease to a stage where there is little prospect of benefit, when the patient may yet be willing to be involved in the research project for the 'collective benefit' that the research will yield. For such research to be ethical, the science must be sound and the alterations induced by the gene transfer therapy should be measurable, and easily assessed within the organ or cells treated, even if there is no discernible clinical benefit.

Initially, it was considered that gene therapy would be used in the management of single gene disorders, such as cystic fibrosis and the inborn errors of metabolism. It was anticipated that children with genetic diseases would be amongst the first candidates for gene therapy. A decision involving children is more complex and usually involves parents or guardians. The foremost consideration must be the best interest of the child. As far as possible, children should be involved in the decision making process. Children who are competent to consent are entitled to do so. In the case of children who are not competent to give consent, valid consent must be obtained from a person with parental responsibility [4,5].

It is a challenge to doctors and researchers to present the information relating to their disease, the alternative treatments and the procedures involved in the research to patients in a way that they can understand. Gene therapy has the added difficulty that the researcher will need to provide details about DNA research, vectors, how genes enter cells and how they function within the cells. There needs to be an extensive ongoing dialogue with the researcher and the patient rather than 'the momentary act of signing a multiple page consent form' [6]. The GTAC has provided precise guidance on the kind of information to be given to patients involved in gene therapy research [7]. This publication gives guidance on the kind of information gene therapy patients should be given, what to include, why the research is being undertaken, confidentiality, and follow-up requirements. A written information leaflet clearly indicating the nature and implications of the research must be provided for the patient. The patient must be given time to absorb the information and have the opportunity to consult an independent counsellor before making a decision to participate in the gene therapy research.

10.4 Confidentiality

Patients agreeing to take part in gene therapy research should be informed that information obtained during the course of the research project will remain confidential and that no publication or report generated from the research would identify the patient. Clear principles govern the confidentiality of personal health service information, the use of that information and the circumstances in which it might be disclosed [8].

10.5 Medical surveillance of gene therapy patients

One of the responsibilities of the GTAC was the 'setting up and maintaining a confidential register of patients who have been the subject of gene therapy' [1]. In obtaining consent of the patient to participate in gene therapy research, consent must also be obtained for the patient to have his/her name placed on a register so that monitoring could be carried out until the patient's death. Indeed, consent must also be obtained to place on the register, the name(s) of any children born to patients who have had gene therapy treatment.

It is inevitable that patients undergoing gene therapy will have adverse experiences related to the gene transfer despite taking every possible precaution to minimize the risks. In addition, the patient may also experience a variety of health problems that may or may not be associated with the gene transfer procedure. Adverse events may possibly occur several years after the gene therapy. Research workers must ensure the long-term follow-up and care of patients who participate in somatic gene therapy. Many complex ethical issues are raised by long-term follow-up, such as the issue of informed consent to follow-up, privacy and confidentiality. Any follow-up system must ensure protection of the patients' rights, while at the same time obtaining information that will provide meaningful results. 'It would be good medical practice to establish a mechanism for tracking patients that would provide the patients and their health care providers with verifiable information concerning gene transfer and guide appropriate diagnostic and therapeutic actions if adverse health events were encountered' [9].

10.6 Germline gene therapy

Germline gene therapy is the alteration of the germline cells, which includes the ova and sperm, as well as the cells from which these are derived. It may also involve manipulation of the zygote. A distinction must be made between interventions that are directly aimed at producing germline effects and the unintended side effects of somatic cell gene therapy. In theory germline cell gene therapy has several advantages compared to somatic cell gene therapy. Germline gene therapy offers the possibility of a 'true' cure for many genetic diseases and by preventing the transmission of disease genes would obviate the need to perform costly somatic gene therapy in many generations. The fear of germline therapy is the risk of inducing unpredictable long-term changes to patients and their children. It is conceivable that therapeutic germline gene therapy would open the door to attempts to enhance human traits, such as intelligence and stature. Genetic modification of human germline cells may have consequences not just for the individual whose cells were originally altered, but also for the individuals who inherit the genetic modification in subsequent generations.

It would inevitably mean a denial of the rights of these individuals to any choice about whether their genetic constitution should have been modified. It is argued that the first generation that would have the power to so deeply shape the next generation should not use it in a way to predestine that next generation's choices [10].

If the scientific barriers are overcome, should germline gene therapy ever be used? Germline gene therapy may be a possible approach to avoiding what would be certain transmission of harmful mutations. The absolute certainty of inheriting a harmful mutation can only occur in two ways. First, a woman who is homoplastic for a harmful mutation in the mitochondrial genome has a 100% risk of having an affected child. Second, a couple, who each have a recessive gene disorder can only have affected children. The majority of harmful gene mutations carry a 50% risk (dominant disorders) or a 25% risk (recessive disorders) of affected offspring. Alternative approaches other than germline gene therapy, such as *in vitro* fertilization and pre-implantation diagnosis provide a simpler way to screen for normal zygotes and select these for implantation, rather than to attempt genetic modification of zygotes identified as having the harmful mutation. There is no significant place for germ-line therapy in the treatment of genetic disease [11]. The Committee on the Ethics of Gene Therapy recommended that gene modification of germline cells should not be attempted [1]. For couples, identified at risk of having a child with a serious genetic disorder, pre-implantation diagnosis and implantation of an unaffected embryo would provide an alternative approach to having a normal child without resorting to germline gene modification. With the recent debate on *in utero* gene therapy, the question of germline gene therapy again has been raised. Some have argued that a total ban on germline therapy is detrimental to research and is hindering progress in this vital field [12].

10.7 *In utero* gene therapy

The prospect of *in utero* gene therapy presents not only tremendous opportunities but also real ethical concerns. In many genetic disorders significant organ damage, especially the brain, will have occurred by the time of birth. Clearly, there are advantages of *in utero* therapy. First, *in utero* therapy would permit the targeting of rapidly expanding populations of stem cells, enabling a better uptake of the vectors and integration of the 'therapeutic' genes. Second, the fetal immune system has not yet fully developed, the fetus is significantly more immune tolerant and would be less likely to develop an immune response against vector and transgene. Third, *in utero* gene therapy would allow 'correction' of the genetic defect before significant organ pathology has occurred. As with somatic gene therapy, *in utero* somatic gene therapy, would be restricted to life-threatening disorders in which significant irreversible organ damage is certain and for which no satisfactory treatment is available. Potential candidate disorders for *in utero* somatic gene therapy include cystic fibrosis, mucopolysaccharidoses, sphingolipid storage diseases, alpha thalassaemia, Duchenne muscular dystrophy and familial intrahepatic cholestasis. However, the risks of *in utero* gene therapy would be much greater than in postnatal somatic gene therapy. There are the potential risks of any invasive *in utero* procedure, such as injury to the fetus, infection, hemorrhage, amniotic fluid leak and preterm labour. However, the main concern would be inadvertent germline alterations due to the somatic transgene finding its way into the recipient's gonads and thus become integrated in the genetic make-up of the individual with transmission to future offspring. Inadvertent germline alteration is not a new ethical concern for it has been a concern in adult somatic gene therapy. Evidence to date would indicate that the overall risk is extremely small [13]. In sheep fetuses, injected intraperitoneally with a retrovirus-mediated neomycin resistant gene, there was no evidence of germline modification of recipient animals. The absence of germline neomycin resistance gene transduction was confirmed after PCR analysis of the sperm of three rams and after screening of lamb offspring bred from primary transduced sheep [14].

The GTAC reported on the potential use of gene therapy *in utero* [15]. It concluded that there were no new ethical issues raised by *in utero* therapy that were not already recognized in other interventions *in utero*, or in the use of gene therapy in other situations. The issue of informed consent remains a matter solely for the pregnant woman. The disease treated would need to be life threatening or associated with severe disability, for treatment after birth is unavailable, in order to justify *in utero* intervention. In the event of a proposal being presented to GTAC the risks of the physical procedures and the potential for germline transmission would need to be fully addressed. Whilst accepting that there may be valid clinical reasons to intervene *in utero* to correct the genetic damage, the report adopted a position that on the basis that any direct *in utero* gene transfer would raise specific ethical and scientific concerns, such approaches would be unacceptable for the foreseeable future. In the light of the potential advantages of *in utero* therapy, preclinical research should be encouraged to identify the appropriate genetic disease, routes of administration, vector system, timing in fetal development for gene transfer and gene expression and outcomes in relevant animal models. It is to be hoped that *in utero* gene therapy in inherited diseases may provide an alternative choice following prenatal diagnosis where at present the only options are termination of the pregnancy or the acceptance of a child with a life-threatening disease or severe disability.

10.8 Conclusion

Human gene therapy is widely acclaimed for its potential to provide novel and powerful therapeutic approaches to the management of inherited and acquired diseases. However, as with all medical advances, gene therapy is associated with ethical concerns. With the rapid pace of advances in molecular medicine, early consideration and anticipation of ethical issues is essential in advance of the new technologies becoming routine clinical practice. The public fears and concerns about gene therapy need to be addressed. Health professionals and scientists have an important role in raising the public awareness about the ethical issues in gene therapy and ensuring that the debate on the ethical concerns is accurately based on fact. In seeking for public understanding and approval, it has to be made clear that the objective of molecular medicine is the alleviation of suffering due to genetic and acquired diseases.

References

1. Report of the Committee on the Ethics of Gene Therapy (1991) HMSO, London.
2. Royal College of Physicians (1990) Research Involving Patients: a working party report. London.
3. Nevin, N.C. (1998) Gene therapy: supervision, obstacles and the future. *Int. J. Pharmaceut. Med.*, **12**, 19–22.
4. Medical Research Council (1991) The Ethical Conduct of Research on Children.
5. General Medical Council (1999) Seeking Patients Consent: The Ethical Considerations, GMC Publications, London.
6. Walters, L. and Palmer, J.G. (1997) *The Ethics of Human Gene Therapy*, Oxford University Press, New York.
7. Gene Therapy Advisory Committee (1994) *Writing an Information Leaflet for Patients Participating in Gene Therapy Research*. Department of Health, London.
8. Confidentiality and Medical Genetics (1998) Genetic Interest Group.
9. Ledely, F.D., Brody, B., Kozinetz, C.A. and Mize, S.G. (1992) The challenge of follow-up for clinical trials of somatic gene therapy. *Hum. Gene Ther.*, **3**, 657–663.
10. Lewis, C.S. (1955) *The Abolition of Man*. MacMillan, New York.
11. Danks, D.M. (1994) Germ-line therapy: no place in treatment of genetic disease. *Hum. Gene Ther.*, **5**, 151–152.
12. Harris, J. (1994) Biotechnology, friend or foe? Ethics and control. In: *Ethics and Biotechnology* (eds A. Dyson and J. Harris). Routledge, London, pp. 216–229.
13. Gordon, J.W. Germline alteration by gene therapy: assessing and reducing the risks. *Mol. Med. Today*, **4**, 468–470.
14. Podara, C.D., Tran, N., Eglitis M., Moen, R.C., Troutman, L., Flake, A.W., Zhao, V., Anderson, W.F. and Zanjani, E.D. (1998) *In utero* gene therapy: transfer and long-term expression of the bacterial neo gene in sheep after direct injection of retroviral vectors into pre-immune sheep. *Hum. Gene Ther.*, **9**, 1571–1585.
15. Report on the Potential Use of Gene Therapy *in Utero* (1998) Gene Therapy Advisory Committee, Health Department of the United Kingdom.

Chapter 11

Prospects for gene therapy

Nicholas R. Lemoine

11.1 Introduction

The field of gene therapy has evolved significantly over the last decade and is now well placed to deliver some of its promise in the clinic. The naivety and hype that characterized the early clinical trials have been replaced by a new sense of realism among both investigators and patients, with recognition that progress will be incremental rather than revolutionary. More effort is now going into establishing robust models for the investigation of mechanisms of gene transfer and expression, key issues of basic biology which were largely ignored at the beginning but are now recognized as absolutely critical to translating gene therapy into the clinic. Animal models of disease have improved dramatically with advances in knockout and transgenic technology, and when combined with the accelerated discovery of disease genes through the genome projects allow meaningful data to be generated in the preclinical phase. The discovery of pluripotent stem cells which can be exploited for the delivery and expression of therapeutic genes in diseased tissues such as dystrophic muscle or tumor neovasculature is a major advance which will find a plethora of therapeutic applications.

11.2 Mechanisms of gene transfer and expression

In retrospect it seems perverse that we should have attempted to deliver therapeutic genes using recombinant viruses without knowledge of how these biological agents recognize and enter cells. However, it is only recently that the coxsackie and adenovirus receptor (CAR) was identified [1], and integrins and receptors for fibroblast growth factors (FGFR) recognized to be involved in the entry of adeno-associated viruses [2,3]. Not only does this new understanding help us to make sense of the previous results, but it gives us the power to manipulate the systems in the ιuture. One of the most exciting prospects now in view is the development of targeted delivery systems exploiting hybrid combinations of viral and synthetic components. For instance, new ligands can be introduced into the coats of recombinant viruses to alter or restrict their tropism [4], while viral proteins can be incorporated into synthetic particles to enhance endosomal release and nuclear localization. Development of such 'smart vectors' will be a major growth area over the next decade.

The huge amount of sequence data generated by the genome projects will lead not only to the identification of new genes but also to full detail of the structure and organization of the genome, including the elements involved in the long-range control and insulation of transcriptional units. In the short term this will help inform the design of therapeutic cassettes for selective and conditional expression in human cells, and in the long term will facilitate the development of mammalian artificial chromosomes [5].

11.3 Development of models of disease and therapy

Construction of transgenic and knockout mice is now a routine procedure which can be achieved in a matter of weeks, and cross-bred strains with multiple genetic abnormalities are easily generated. The ability to introduce human genes into the germline of these animals allows us to get closer to models which accurately reproduce the human condition, and the extension of transgenic technology to other species adds further power to the approach. Work in animal models has shown that there are stem cells in various tissues including bone marrow (but also, surprisingly, brain [6]) which have the capacity for populating other organs and tissues where they can show differentiation in surprising directions, taking on the characteristics of muscle cells or activated endothelium for instance. These stem cells can now be isolated with relative ease and they can be genetically manipulated with a variety of vector systems. Their plasticity of development and potential homing to diseased organs will allow investigators to exploit them for cell-based gene transfer, and this approach is likely to be translated to the clinic for a range of different disease applications.

11.4 Cycles of clinical development – beyond the phase I trial

It is clear that the classical approach of escalating amounts of drug to maximal tolerated dose in phase I trials in end-stage patients is largely inappropriate to testing the principle (and safety) of gene therapy in humans. Indeed, for such approaches as genetic vaccination escalating the dose may actually be counterproductive to therapeutic effect. It is instructive that only some of the prototype gene therapeutics, such as the selectively replicating E1B-deleted adenovirus and p53 replacement vectors for cancer therapy, have successfully passed through phase I/II trials, are poised for phase III testing and may ultimately reach the market. Future trials will need to be designed in a more sophisticated way, incorporating molecular pathology and functional imaging technologies to validate and quantitate the consequences of genetic intervention. The application of such approaches as laser capture microscopy [7] to isolate individual cell populations from clinical specimens and DNA microarray technology for gene expression profiling [8] has begun to change the way in which conventional small molecule drugs are assessed, and must be adopted for analysis of gene therapeutics. Advances in functional imaging by positron emission tomography (PET) scanning and magnetic resonance imaging (MRI) for so-called 'metabonomics' look very promising for the development of surrogate markers of gene expression [9,10].

It will be important for gene therapeutics to pass through cycles of development, being tested in preclinical models and clinical series before returning to the laboratory to incorporate advances in gene delivery technology, transcriptional control and immunological manipulation. The evolution of the agents presently available has been impressive but it is evident that much yet needs to be done in the platform technologies before the optimal system emerges for any one application.

The future for gene therapy looks bright. We can expect the field to advance for the benefit of patients as scientists and clinicians use their imagination and work together to make gene therapy a clinical reality. Those who understand where we are now will be those who can take up the challenges for the future.

References

1. Bergelson, J.M., Cunningham, J.A., Droguett, G., Kurt-Jones, E.A., Krithivas, A., Hong, J.S., Horwitz, M.S., Crowell, R.L. and Finberg, R.W. (1997) Isolation of a common receptor for Coxsackie B viruses and adenoviruses 2 and 5. *Science*, **275**, 1320–1323.
2. Summerford, C., Bartlett, J.S. and Samulski, R.J. (1999) AlphaVbeta5 integrin: a co-receptor for adeno-associated virus type 2 infection. *Nat. Med.*, **5**, 78–82.
3. Qing, K., Mah, C., Hansen, J., Zhou, S., Dwarki, V. and Srivastava, A. (1999) Human fibroblast growth factor receptor 1 is a co-receptor for infection by adeno-associated virus 2. *Nat. Med.*, **5**, 71–77.
4. Dmitriev, I., Krasnykh, V., Miller, C.R., Wang, M., Kashentseva, E., Mikheeva, G., Belousova, N. and Curiel, D.T. (1998) An adenovirus vector with genetically modified fibers demonstrates expanded tropism via utilization of a coxsackievirus and adenovirus receptor-independent cell entry mechanism. *J. Virol.*, **72**, 9706–9713.
5. Vos, J.M. (1998) Mammalian artificial chromosomes as tools for gene therapy. *Curr. Op. Genet. Dev.*, **8**, 351–359.
6. Vescovi, A.L., Parati, E.A., Gritti, A. *et al.* (1999) Isolation and cloning of multipotential stem cells from the embryonic human CNS and establishment of transplantable human neural stem cell lines by epigenetic stimulation. *Exp. Neurol.*, **156**, 71–83.

7. Sirivatanauksorn, Y., Drury, R., Crnogorac-Jurcevic, T., Sirivatanauksorn, V. and Lemoine, N.R. (1999) Laser-assisted microdissection: applications in molecular pathology. *J. Pathol.* (in press).
8. Various authors (1999). *Nat. Genet.*, **21** (Suppl. 1).
9. Gambhir, S.S., Barrio, J.R., Phelps, M.E. *et al.* (1999) Imaging adenoviral-directed reporter gene expression in living animals with positron emission tomography. *Proc. Natl Acad. Sci. USA*, **96**, 2333–2338.
10. Tjuvajev, J.G., Avril, N., Oku, T. *et al.* (1998) Imaging herpes virus thymidine kinase gene transfer and expression by positron emission tomography. *Can. Res.*, **58**, 4333–4341.

Index

Adeno-associated viral vectors,
   drawbacks of, 6
Adeno-associated virus,
   cell entry receptors, 163
   delivery of suicide genes, 17
   genome, 33
   helper virus, 33
   limitations, 34
   titre, 34
   transduction efficiencies, 33
   use in targeted integration, *see*
      Gene delivery targeting
Adenosine deaminase deficiency
   (ADA-SCID),
   ADA, 13, 71
   clinical features, 13, 73
   clinical trials, 13, 76–77, 149–150
   current therapies, 13, 73
   pre-clinical trials, 74–76
   use of retroviral vectors in *ex
      vivo* gene therapy, 72–74, 125–
      126
Adenoviral vectors,
   advantages of 5, 29–30
   cancer, 5
   cystic fibrosis, 5, 151
   gutless or pseudoadenovirus
      (PAV) adenoviruses, 5, 32
   inflammatory response, 5
   obstacles, 29–30
   oral vaccines, 30
   packaging size of DNA, 32
   replication control, 5
   targeting, 32, 132–133

Adenovirus,
   cell surface receptors, 27
   early and late genes, 27–29
   genome, 27–29, 30
   gutless, 138
   infection, 132
   inflammatory response to, 30–31
   inverted terminal repeats (ITRs),
      29
   less-immunogenic viruses, 31–32
   packaging sequences, 29, 32
   replication competent
      adenovirus (RCA), 31
   structure, 27
   titres, 29
   transcription of viral genes, 2
   types, 27
   use in endosome escape, 135
Amyotrophic lateral sclerosis, 152
Angiogenesis,
   strategies for reduction of,
      91–92
   anti–angiogenic agents, 92
Animal models, 4, 164
Antisense oligonucleotides, 15
Antisense therapy, 15, 16, 89, 106
Antiviral gene therapy,
   application to HIV infection,
      100, 104–113, 151
   clinical trials, 151–152
   strategies of, 100–101
Asialoglycoprotein receptor, 130
Atherosclerosis,
   genetic mechanisms, 93

gene therapy for, 94–95
treatment of, 93–94

Bystander effect, 16, 138

Cancer,
    antisense therapy, 15
    breast, 14, 15, 19, 127, 137
    choriocarcinoma, 14
    colorectal, 14, 19, 83–88, 127,
        137
    ERBB2, 15, 89, 91, 136, 137
    gene therapy strategies, 15–16,
        89–92, 127
    germ cell, 14
    glioblastomas, 128
    glioma, 4, 56
    hepatoma, 50, 134–135
    leukemia, 14, 15
    lung, 14, 15, 19, 137
    lymphoma, 14, 19, 134
    melanoma 16, 19, 127, 137, 147
    myeloma, 19
    ovarian, 14, 19
    pancreatic, 15 137
    p53, 15
    RAS, 15, 89, 91
    retinoblastoma, 15
Cell targeting,
    engineering of retroviral
        envelopes, 130–132
    natural vector tropism, 128–129
    use of adenoviral vectors, 32,
        132–133
    use of retroviral pseudotypes,
        129–130
    targeting moieties, 50, 54–55,
        130–135
Cystic fibrosis,
    adeno-associated gene delivery,
        82
    adenoviral gene delivery, 3, 13,
        79, 80–81
    clinical features, 12–13, 78
    clinical studies, 80, 151

cystic fibrosis transmembrane
        conductance regulator, 3, 13,
        78, 151
    gene, 3, 13, 78
    liposome gene therapy, *see* Non-
        viral vectors
    pre-clinical studies, 79–80
    treatment of, 13
Cytokines, 90, 127

Diabetes mellitus,
    ex vivo gene therapy of, 96–97
    insulin gene, 94, 96
    in vivo gene therapy of, 97
    types, 94
Duchenne muscular dystrophy, 8

Epidermal growth factor (EGF)
    receptor, 130–132

Familial hypercholesterolemia,
    clinical features, 14
    clinical trials, 14, 150–151
    LDL receptor, 14, 150
    treatment of, 14

Gene delivery,
    bone marrow, 74, 127, 134
    brain, 55, 128
    endothelial cells, 50, 127
    hematopoietic cells, 26, 35, 72,
        74, 75, 126
    hepatocytes, 26, 36, 130, 133–135
    hepatoma, 50, 130, 135
    keratinocytes, 127
    lung, 55
    lymphocytes, 72, 75, 76, 126
    myoblasts, 34
    myotubes, 34
    neurons, 26
    skin fibroblasts, 127
    tracheal and bronchiolar
        epithelium, 51–52, 151
Gene delivery systems,
    cationic lipids, *see* Non-viral
        vectors

electroporation, *see* Non-viral vectors
gene gun, 8, 59–60
liposomes, *see* Non-viral vectors
receptor-mediated gene transfer, *see* Receptor mediated gene transfer viral, *see* Adenovirus, Adeno–associated virus, Herpes simplex virus, Retrovirus
virosomes, *see* Virosomes
Gene delivery targeting,
cell targeting, *see* Cell targeting,
ex vivo gene delivery, *see* Adenosine deaminase deficiency
naked DNA, *see* Non-viral vectors
non-viral vectors, *see* Non-viral vectors
targeted integration, 129
transcriptional targeting, 90, 135–138
transgene, 90, 138
Gene therapy,
clinical trials, 9, 145–152
definition, 1, 11, 71
ethical and regulatory considerations, 155–162, *see also* Germline gene therapy
future of, 19, 164–165
ganciclovir, 16, 121, 128
genetic prodrug activation therapy (GPAT), *see* Suicide gene therapy
graft-versus-host disease (GVHD), 149
goal, 1, 22
immunogenicity to gene therapy constructs, 121
industrial areas of concern, 17–18
intended use, 1
in utero, 160–161

obstacles 17, 121, 125
potential, 9
somatic, *see* Somatic gene therapy
suicide gene, 4, 17
target diseases, 1, 8, 19, 83
tissue/tumor specific promoters, 90, 135–137
vaccines, *see* Vaccines
Gene therapy trials,
classification of studies, 143, 145
gene marking studies, 145–148
of cancer, 148–149
of infectious diseases, 151
of monogenic diseases, 149–151
overview of vectors and routes of administration, 143–145
safety, 148
worldwide overview of protocols, 142
Gene Therapy Advisory Committee (GTAC), 156–160
Germline gene therapy, 1, 159–160

Hemophilia B,
gene therapy of, 126–127
Hematopoietic stem cells, 26, 72, 114, 126
Herpesvirus
cell tropism, 34, 128
genome, 34
structure, 34
target cells, 34
titre, 34
Herpes simplex virus thymidine kinase, 4, 15, 120–121, 134, 137
Herpes simplex viral vectors,
construction, 34
delivery to myoblasts and myotubes, 34
disabled infectious single copy virus, 6
gutless vector, 35
potential use in gene therapy, 34
treatment of glioblastoma, 128

Human immunodeficiency virus (HIV) anti-HIV gene therapy, *see* Antiviral gene therapy cell surface receptor, 102–103, 111
genome, 102–103
immune response to, 103–104
infection, 16, 111
life cycle, 100–102
pathogenic retrovirus, 4, 24, 99
RNA decoys, 108–109
treatments, 16–17, 106

Immune system,
antigen recognition, 115
dendritic cells, 91
effector cells, 114–115
major histocompatability complex (MHC), 114–116
roles, 114
vaccines, *see* Vaccines
Immunotherapy, 16, 90–91, 113–122, 148
Infectious diseases, 16
antiviral gene therapy, *see* Antiviral gene therapy
conventional treatment, 99
manipulation of immune response, 113–122

Laser capture microscopy, 165
Liposomes, *see* Non-viral vectors
Locus control regions (LCR), 135

Magnetic resonance imaging (MRI), 165
Matrixmetalloproteases, 88
Microarray technology, 165
Multiplicity of infection (MOI), 37
Multidrug resistance genes, 91

Non-viral vectors,
biolistics, 59
cationic liposome, 7, 43, 45–51
cationic polymer, 43, 52–57
chimeric proteins, 57–58

cytofectin, 45, 46
electroporation, 59
endosome escape, 50–51, 54,55
formulation of vesicles, 46
gene delivery for cystic fibrosis, 13, 46–49, 80–81, 151
hybrid lipoplex systems, 51
immunoliposomes, 134
lipoplex, 43
mechanism of entry into a cell, 44
micro-injection, 59
naked DNA, 4, 6–8
polyplex, 44
principles, 43–44
problems with, 45, 49, 133
Sendai virus (HVJ), *see* Viral vectors
stealth liposomes, 133
steric stabilization, 49–50
targeting, 50, 54–55, 133–134
VP22, 4, 8

Oligonucleotides, *see* Antisense oligonucleotides; Antisense Therapy

Plasminogen, 88
Positron emission tomography (PET) scanning, 165
Promoters,
associated α-fetoprotein (AFP), 136
carcinoembryonic antigen (CEA), 137
ERBB2, 137
summary table, 136
tyrosinase, 137

Ram-1, 131
Receptor-mediated gene transfer,
asialoglycoprotein receptor, 130, 133
asialoorosomucoid receptor (ASOR), 135

EGF receptor, 130–132
galactose receptor, 50
gastrin-releasing peptide (GRP), 133
lactose receptor, 134
problems with, 135
ram-1 receptor, 131
transferrin receptor, 54, 134
Reticuloendothelial system (RES), 133
Retroviral vectors,
    construction, 25
    delivery of cytokine genes, 127
    delivery of HIV coat proteins, 17, 151
    drawbacks, 26
    Moloney murine leukemia virus (MoMuLV), 25
    natural vector tropism, *see* Cell targeting
    somatic gene transfer, *see* Somatic gene therapy,
    suicide gene therapy, *see* Suicide gene therapy
    targeting ligands, *see* Cell targeting
    use in targeted integration, *see* Gene delivery targeting
    use in *ex–vivo* approach, 25, 125–127
Retrovirus(es),
    core genes, 23
    genome structure, 24
    infection, 5
    integration, 5, 23
    packaging, 22
    problems with, 5, 25
    pseudotypes, 129
    replication-competent retroviruses (RVR), 26
    replication incompetent, 5
    structure, 22
    titre, 25
Ribozymes, 106–108
RNA decoys, *see* Human immunodeficiency virus (HIV)

Rous sarcoma virus (RSV), 24, 129

Somatic gene therapy,
    clinical trials, 146
    definition, 2
    divisions of, 2–4
    retroviral delivery, 25
Suicide genes, 4, 17, 148
Suicide gene therapy, 15, 89–90, 135–138, 148–149

Tissue inhibitors of matrix metalloproteinases (TIMPs), 92
Tumor,
    immunogenicity, enhancement of, 90
    specific promoters, 136
    suppressor genes, 15, 89, 93
    targeting, *see* Cell Targeting

Vaccines,
    cytomegalovirus (CMV), 120–121
    gene therapy, 117, 127
    genetically altered immune cells, 120–121, 127
    HIV, 117–119, 122
    influenza A, 118
    protective, 115–119
    therapeutic, 119–120
Vascular endothelial growth factor (VEGF), 88, 92
Viral Vectors,
    adeno-associated virus
    adenovirus 5, *see also* Adenoviral vectors
    avian leukosis virus, 131
    baculoviruses, 36
    drawbacks of, 6
    Epstein-Barr virus, 37
    feline immunodeficiency virus (FIV), 26
    human spumaviruses, 5
    hybrid vectors, 35–37
    lentiviruses, 26, 36

major properties of, 24
murine leukemia virus, 5, 130–132
Onyx-virus, 18, 89
other viral vectors, 35
parvoviruses, 128
retrovirus, *see* Retrovirus(es)

Sendai virus (hemagglutinating virus of Japan [HVJ]), 51–52, 134
Virosomes, 134

Xenografts, 56